THE COMMONWEALTH AND INTERNATIONAL LIBRARY
Joint Chairmen of the Honorary Editorial Advisory Board
SIR ROBERT ROBINSON, O.M., F.R.S., LONDON
DEAN ATHELSTAN SPILHAUS, MINNESOTA

MECHANICAL ENGINEERING DIVISION
General Editor: N. HILLER

STRESS ANALYSIS
PROBLEMS IN S.I. UNITS

STRESS ANALYSIS PROBLEMS IN S.I. UNITS

BY

D. F. MALLOWS, B.Sc. (Hons.)
*Senior Lecturer in Mechanical Engineering
Oxford Polytechnic*

AND

W. J. PICKERING, B.Sc.
*Lecturer in Mechanical Engineering
Oxford Polytechnic*

PERGAMON PRESS
OXFORD · NEW YORK
TORONTO · SYDNEY · BRAUNSCHWEIG

Pergamon Press Ltd., Headington Hill Hall, Oxford
Pergamon Press Inc., Maxwell House, Fairview Park, Elmsford, New York 10523
Pergamon of Canada Ltd., 207 Queen's Quay West, Toronto 1
Pergamon Press (Aust.) Pty. Ltd., 19a Boundary Street, Rushcutters Bay, N.S.W. 2011, Australia
Vieweg & Sohn GmbH, Burgplatz 1, Braunschweig

Copyright © 1972 D. F. Mallows and W. J. Pickering

All Rights Reserved. No part of this publication may be reproduced, stored in a retrieval system, or transmitted, in any form or by any means, electronic, mechanical, photocopying, recording or otherwise, without the prior permission of Pergamon Press Ltd.

First edition 1972
Library of Congress Catalog Card No. 71-171 465

Printed in Hungary

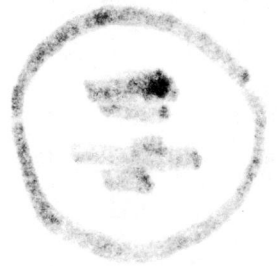

This book is sold subject to the condition
that it shall not, by way of trade, be lent,
resold, hired out, or otherwise disposed
of without the publisher's consent,
in any form of binding or cover
other than that in which
it is published.

08 016292 4 (flexicover)
08 016293 2 (hard cover)

Contents

Preface vii

Introduction viii

1 Statics 1

Equilibrium conditions. Free-body diagrams. Applications to beam reactions and to solutions of statically determinate plane and space frames.

2 Stress and Strain 34

Elastic direct and shear stresses and strains. Elastic moduli E, G, and K. Poisson's ratio. Direct stresses in three-dimensional systems. Relation between E, K, and G. Strain energy in tension and compression. Suddenly applied loads. Stresses and strains in pressurized thin cylinders, compound bars, thermal stresses.

3 Two-dimensional Stress Systems 60

Complementary shear stresses. Effect of direct tensile and compressive stresses on oblique planes. General two-dimensional stress systems. Mohr's circle for stress. Principal stresses.

4 Stresses in Beams 81

Bending moments and shearing forces. Bending stresses. Shear-stress distribution. Combined bending and direct stress. Eccentric loading.

5 Torsion 112

Solid and hollow cylindrical shafts. Power transmission. Keyed and flanged couplings. Combined bending and torsion.

6 Strain Analysis 131
Mohr's circle for strain. The strain rosette. Relation between E, G, and v.

7 Beam Deflections 150
Integration of the elastic line. Macaulay's method. Moment-area method. Superposition principle. Fixed beams.

8 Strain Energy Methods 177
Castigliano's first and second theorems. Deflections of beams and shafts. Determinate and indeterminate plane trusses.

9 Elementary Plastic Stress Analysis 202

10 Analysis of Stress in Engineering Components 224
Close- and open-coiled springs. Leaf springs. Composite beams including reinforced concrete. Thick cylinders. Struts. Theories of failure.

Appendix I S.I. Units Used in this Book 261

Appendix II Properties of Common Materials 262

Index 263

Preface

THIS book has been written to cover topics usually dealt with in HNC and HND strength of materials subjects, in CEI Part I, in the London degree subject properties of materials and stress analysis, and it should also be of use to students studying for CNAA degrees.

With the change to S.I., students and lecturers need to achieve facility in "thinking metric", and this is not attained while it is continually necessary to convert from Imperial to S.I. units. Problems—original and drawn from past examination papers—have therefore been rewritten in S.I. units, numerical values being rounded to achieve rational metric sizes.

The authors wish to express their thanks to the Senate of the University of London for permission to use questions from their examinations. Where changes have taken place to the questions in the process of adapting them to S.I., the responsibility is entirely that of the authors; nor is the university committed to approve the solutions or answers given.

Introduction

THE International System of Units (Système International d'Unités) (S.I.) has, as one of its main advantages, the property of being coherent; that is the unit of a derived quantity is always formed as a product or quotient of two or more of the S.I. base units. A list of the base and derived units used in this book is given in Appendix I.

The recommendations of BSI in PD 5686, 2nd edition, have been followed and found, in the opinion of the authors, to be very satisfactory. In particular, the data in problems and answers has been quoted in the most convenient recommended multiple or sub-multiple of the S.I. unit, but only the units themselves have been used in calculations; thus the diameter of a rod may be quoted as 7 mm, but in calculations this dimension will be in the form 7×10^{-3} m.

In the stress analysis of a system supporting the weight of certain masses, the weights have been quoted in the appropriate unit—the newton. An alternative would be to give the mass in kilograms together with the local value of gravitational acceleration and leave the student to deduce the weight from the relation $W = Mg$. The alternative was not used because it merely adds another step of numerical calculation without contributing anything to the understanding of the stress analysis.

CHAPTER 1

Statics

Definitions and Theory

(a) A free-body diagram is a diagram of the whole, or a part, of a body free from all other objects with which it is, in fact, in contact. Forces or moments transmitted to the free-body by its surroundings are shown on the diagram as external forces or moments.

(b) A body will be in static equilibrium, i.e. it will remain without translational or rotational motion if both the vector sum of the forces is zero and the vector sum of the moments is zero.

(c) Forces and moments are both vector quantities: they can be represented in magnitude and direction by a directed line of suitable length and orientation and can be added graphically by vector triangles, parallelograms, or polygons.

(d) Bow's notation is a method of naming force vectors in vector polygons. A capital letter is written in each space between forces and the force is then referred to by the letters which lie on either side of it.

(e) A system of forces is said to be "statically determinate" if all the forces can be determined by the application of methods based on the above outline. If there are too many unknown forces in the system it will be impossible to find the values of all of them by the methods of statics, and the system is then said to be "statically indeterminate".

Worked Examples

1.1. Draw a free-body diagram of an internal combustion engine piston showing and naming the chief forces acting upon it at the moment when the crank is (a) 30° before the end of the compression stroke, and (b) 30° after the beginning of the power stroke.

The solution is given in Fig. 1.1.

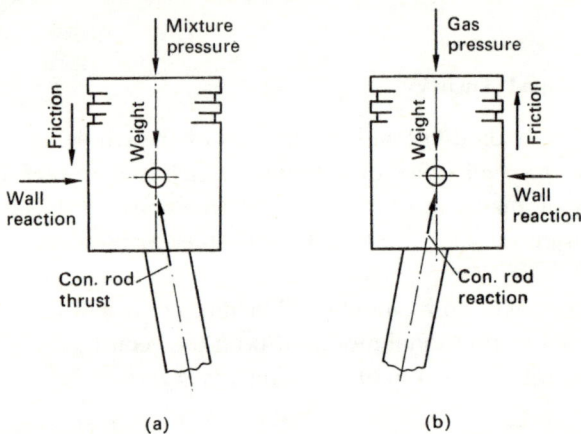

Fig. 1.1.

1.2. Draw free-body diagrams of two smooth cylinders resting in a smooth V-groove, 40 mm wide at the bottom, whose sides are inclined at $22\frac{1}{2}°$ on either side of the vertical. The lower cylinder is 50 mm diameter and the higher 25 mm diameter. Treat the cylinders (a) as two separate bodies, and (b) as a single unit.

The solution is given in Fig. 1.2.

1.3. Figure 1.3 shows a nut and bolt fitted to the free end of a rigidly mounted horizontal cantilever. To loosen the nut a fitter tries unsuccessfully the following methods:

(a) He applies equal horizontal forces of 85 N to each handle of a T-spanner such that each force acts at 150 mm from the spanner centre line.

STATICS

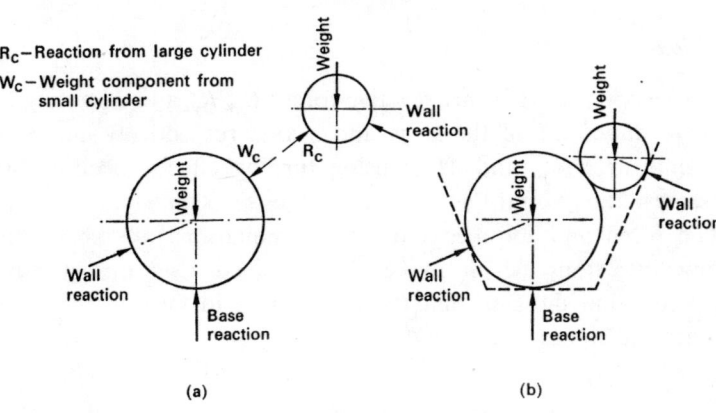

Fig. 1.2.

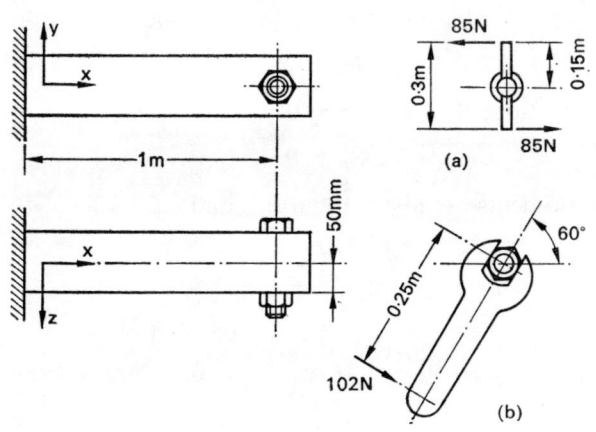

Fig. 1.3.

(b) He applies a force of 102 N perpendicular to the handle of a straight set-spanner at 250 mm from the centre of the nut. The spanner is at 60° to the horizontal.

For each method, find the components of force and moment reactions at the wall, relative to the x, y, z axes shown, caused by the forces applied to the spanners.

Solution

It is assumed that there are reactive forces R_x, R_y, and R_z acting in the positive directions of the x, y, and z axes respectively and reactive moments M_x, M_y, and M_z causing turning effects, positive by the corkscrew rule, about the x, y, and z axes respectively.

The force and moment equilibrium equations are now applied, taking directions which make the assumed reactions positive, a negative value showing that the reaction is, in fact, in the opposite direction to the assumption.

Method (a)

Since there are no applied forces in the y and z directions,
$$R_y = R_z = 0.$$
In the x direction, $R_x + 85 - 85 = 0$,
$$R_x = 0.$$
The only moments are about the z axis and
$$M_x = M_y = 0,$$
$$M_z - 85 \times 0 \cdot 3 = 0,$$
$$M_z = 25 \cdot 5 \text{ N m}.$$
$[R_x = R_y = R_z = 0. \quad M_x = M_y = 0. \quad M_z = 25 \cdot 5 \text{ N m}.]$

Method (b)

The force on the spanner can be replaced by a force, acting at the centre of the nut, of 102 N at 30° to the horizontal together with a couple of 25·5 N m positive with respect to the z axis.

Equilibrium of forces:
In x direction:
$$R_x + 102 \cos 30 = 0,$$
$$R_x = -88 \cdot 3 \text{ N}.$$

In y direction:
$$R_y - 102 \sin 30 = 0,$$
$$R_y = 51 \text{ N}.$$

In z direction there are no forces:
$$R_z = 0.$$
$[R_x = -88\cdot3 \text{ N}. \quad R_y = 51 \text{ N}. \quad R_z = 0.]$

Since the nut is not on the centre line of the cantilever, the force on the spanner may have moments about all three axes.

Equilibrium of moments:

About x axis:
$$M_x + 102 \sin 30° \times 0\cdot05 = 0,$$
$$M_x = -51 \times 0\cdot05,$$
$$M_x = -2\cdot55 \text{ N m}.$$

About y axis:
$$M_y + 102 \cos 30° \times 0\cdot05 = 0,$$
$$M_y = -88\cdot3 \times 0\cdot05,$$
$$M_y = -4\cdot415 \text{ N m}.$$

About z axis:
$$M_z - 102 \sin 30° \times 1 + 25\cdot5 = 0,$$
$$M_z - 51 + 25\cdot5 = 0,$$
$$M_z = 25\cdot5 \text{ N m}.$$
$[M_x = -2\cdot55 \text{ N m}. \quad M_y = -4\cdot415 \text{ N m}. \quad M_z = 25\cdot5 \text{ N m}.]$

1.4. A horizontal beam of negligible weight rests on two supports A and B a distance L apart. The left-hand support is a simple roller and the right-hand support is a pin. A moment of magnitude M is applied to the beam around the right-hand support. Find the support reactions.

Solution

Refer to Fig. 1.4.

Assume upward reactions R_A and R_B at the supports.

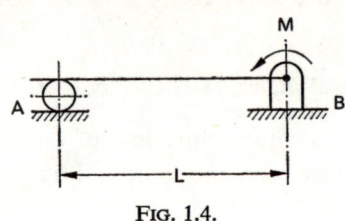

Fig. 1.4.

Apply the moment equilibrium equation $\Sigma M = 0$ about support B taking counter-clockwise moments as positive.

$$M - R_A L = 0,$$

$$R_A = \frac{M}{L}.$$

Apply the force equilibrium equation $\Sigma F = 0$ taking upward forces as positive.

$$\frac{M}{L} + R_B = 0,$$

$$R_B = -\frac{M}{L}.$$

[Reactions are: Upward force M/L at A. Downward force M/L at B.]

1.5. A symmetrical three-hinged arch has a span of 25 m and a rise to the centre pin of 7·5 m. Both supports are on the same horizontal level and a vertical load of 240 kN is applied to the arch at 5 m from the left-hand support. Find the horizontal and vertical components of the forces at each support. [London Univ., B.Sc. 1]

Solution

In a three-hinged arch the hinges are all assumed to be frictionless pins and therefore no turning moment can be transmitted across any pin. The equilibrium equations can be written for the arch as a whole and for each half considered as a separate free-body.

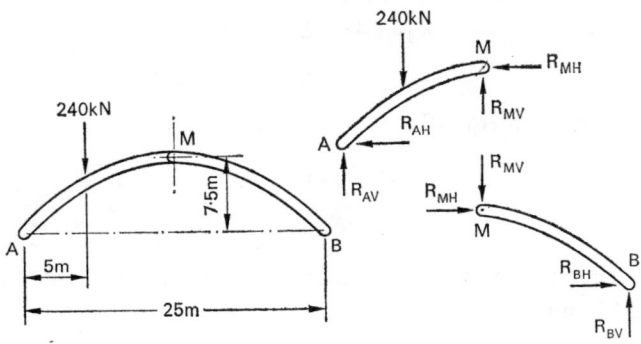

Fig. 1.5.

The directions of the reactions are assumed to be as shown in Fig. 1.5. Applying equilibrium equations to section AM of the arch, sum of moments about $A = 0$. Counter-clockwise positive.

$$-240 \times 5 + R_{MH} \times 7 \cdot 5 + R_{MV} \times 12 \cdot 5 = 0. \tag{1}$$

Sum of vertical forces $= 0$ positive upwards,

$$R_{AV} + R_{MV} - 240 = 0. \tag{2}$$

Sum of horizontal forces $= 0$ positive to right,

$$-R_{AH} - R_{MH} = 0. \tag{3}$$

Applying equilibrium equations to BM:

Sum of moments about $B = 0$ counter-clockwise positive.

$$R_{MV} \times 12 \cdot 5 - R_{MH} \times 7 \cdot 5 = 0. \tag{4}$$

$$\therefore R_{MV} = \tfrac{3}{5} R_{MH}.$$

Substituting this value in eqn. (1),

$$-1200 + 7\cdot5 R_{MH} + 12\cdot5 \times \tfrac{3}{5} R_{MH} = 0.$$

$$R_{MH} = 80 \text{ kN}.$$

$$\therefore R_{MV} = \tfrac{3}{5} \times 80 = 48 \text{ kN}.$$

Substituting this value in eqn. (2),

$$R_{AV} + 48 - 240 = 0,$$

$$R_{AV} = 192 \text{ kN},$$

and substituting in eqn. (3),

$$-R_{AH} - 80 = 0,$$

$$R_{AH} = -80 \text{ kN}.$$

[$\therefore$ reactions at A are: Horizontal 80 kN to the right. Vertical 192 kN upwards.]

Sum of vertical forces on section $BM = 0$. Positive upwards.

$$R_{BV} - R_{MV} = 0.$$

$$\therefore R_{BV} = R_{MV}.$$

$$\therefore R_{BV} = 48 \text{ kN}.$$

Sum of horizontal forces on section $BM = 0$. Positive to right.

$$R_{MH} + R_{BH} = 0.$$

$$\therefore R_{BH} = -R_{MH}$$

$$= -80 \text{ kN}.$$

[Reactions at B are: Horizontal 80 kN to the left. Vertical 48 kN upwards.]

1.6. The plane frame drawn in Fig. 1.6 is pinned to a rigid wall at joints 1 and 2; it supports a vertical load of 100 kN at joint 4. Assuming all joints are frictionless pins, find the forces in all members.

STATICS

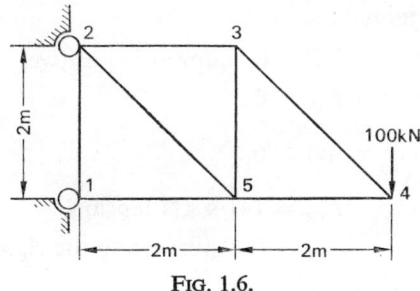

FIG. 1.6.

Solution

This solution uses the method of equilibrium of the joints, a satisfactory method for many problems especially where the forces in all members are required and where the angles can easily be found. Assuming all unknown forces are tensile:

Equilibrium of joint 4:

$$\Sigma F_V = 0 \quad \text{upward forces positive,}$$

$-100 + F_{34} \sin 45° = 0,$

$$F_{34} = 100\sqrt{2} = 141\cdot 4 \text{ kN tension.}$$
$$\Sigma F_H = 0 \quad \text{positive to the right,}$$

$-F_{54} - F_{34} \cos 45° = 0,$

$$F_{54} = -100\sqrt{2}/\sqrt{2} = 100 \text{ kN compression.}$$

Equilibrium of joint 3:

$$\Sigma F_V = 0 \quad \text{upward forces positive,}$$

$-F_{35} - F_{34} \sin 45° = 0,$

$$F_{35} = -100 = 100 \text{ kN compression.}$$
$$\Sigma F_H = 0 \quad \text{positive to the right.}$$

$F_{34} \cos 45° - F_{23} = 0,$

$$F_{23} = 100 \text{ kN tension.}$$

Equilibrium of joint 5:

$$\Sigma F_V = 0 \quad \text{upwards positive,}$$
$$F_{25} \sin 45° + F_{35} = 0,$$
$$F_{25} \times \frac{1}{\sqrt{2}} - 100 = 0,$$
$$F_{25} = 141\cdot 4 \text{ kN tension.}$$
$$\Sigma F_H = 0 \quad \text{positive to the right.}$$
$$F_{54} - F_{25} \cos 45° - F_{15} = 0,$$
$$-100 - 141\cdot 4 \times \frac{1}{\sqrt{2}} - F_{15} = 0,$$
$$F_{15} = -200 \text{ kN} = 200 \text{ kN compression.}$$

1.7. In the pin-jointed truss shown in Fig. 1.7 the joints on the line AG are all 2 m apart. The vertical members FH and BM are 1 m,

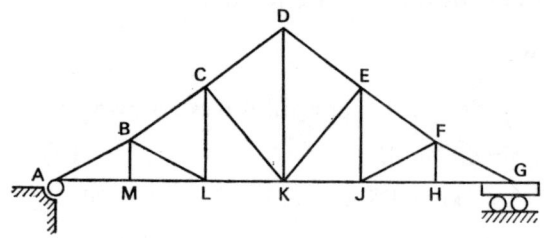

Fig. 1.7.

CL and EJ are 2·5 m, and KD is 4 m long. Loads of 4 kN act vertically down at each of the joints B, C, D, E, and F, and loads of 1 kN act vertically down at each of the joints M, L, K, J, and H. Find, analytically, the loads carried by members CD, CK, and CL.

[London Univ. B.Sc., 1956]

Solution

Since the frame and loading are both symmetrical about the vertical centre line the reactions at each end are equal to half the total load,

$$R_A = R_G = 12\cdot 5 \text{ kN.}$$

In problems where the loads are required in only a few members, the method of sections (adopted here) is often the simplest.

Assume there is tension in all members in which the forces are unknown and consider the equilibrium of the section of frame to the left of a vertical line just to the right of member CL.

The forces in CD, CK, and KL are all unknown, but taking moments about point K will eliminate two of them.

Taking counter-clockwise moments as positive,

$$-6R_A + 5 \times 4 + 5 \times 2 - 2 \cdot 5 F_{CD} \times \frac{2}{\sqrt{(2^2 + 1 \cdot 5^2)}} - 2 F_{CD} \frac{1 \cdot 5}{\sqrt{(2^2 + 1 \cdot 5^2)}} = 0.$$

$$\therefore F_{CD} = -14 \cdot 07 \text{ kN},$$
$$F_{CD} = 14 \cdot 07 \text{ kN compression.}$$

Consider vertical equilibrium of same section with upward forces positive:

$$12 \cdot 5 - 5 - 5 - 14 \cdot 07 \times \frac{1 \cdot 5}{2 \cdot 5} - F_{CK} \frac{2 \cdot 5}{3 \cdot 202} = 0,$$

$$F_{CK} = -7 \cdot 63 \text{ kN},$$
$$F_{CK} = 7 \cdot 63 \text{ kN compression.}$$

To find force F_{CL} it is simplest to use the equilibrium of joint C:

$\Sigma F_V = 0$ upward forces positive,

$$-14 \cdot 07 \times \frac{1 \cdot 5}{2 \cdot 5} - F_{BC} \times \frac{1 \cdot 5}{2 \cdot 5} - F_{CL} + 7 \cdot 63 \times \frac{2 \cdot 5}{3 \cdot 202} = 0.$$

$\Sigma F_H = 0$ positive forces to the right,

$$-14 \cdot 07 \times \frac{2}{2 \cdot 5} - F_{BC} \times \frac{2}{2 \cdot 5} - 7 \cdot 63 \times \frac{2}{3 \cdot 202} = 0,$$

$$F_{BC} = -20 \cdot 02 \text{ kN},$$
$$F_{CL} = 9 \cdot 53 \text{ kN tension.}$$

1.8. The pin-jointed plane frame shown in Fig. 1.8 is attached to a vertical wall by pin supports at A and B. Note that the two members

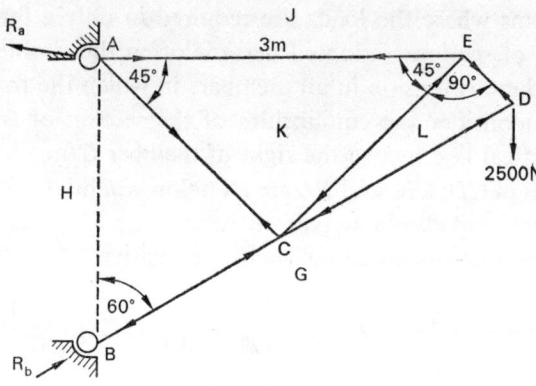

FIG. 1.8.

BC and *CD* lie in the same line. Find (a) the magnitude and type of reactions at the supports, and (b) the magnitude and type of the force in every member.

Solution

The problem is most easily solved graphically by first drawing the frame to scale and then applying the equilibrium equations in vector form, i.e. $\Sigma \mathbf{F} = 0$ by drawing a closed force polygon.

To find the reactions, notice that, since member *BC* is pin-jointed

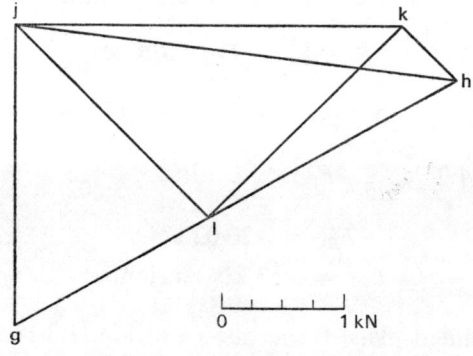

FIG. 1.9.

at both ends and forces are only applied to it at the joints, the force in *BC* can only be along its line. Thus, the reaction at *B* must be along *BC*. The frame as a whole is in equilibrium under three non-parallel forces, and their lines of action must therefore meet at a point. This point must be joint *D*. The reaction at *A* is therefore along line *DA*.

The space diagram can then be lettered using Bow's notation, and the triangle *jgh* in Fig. 1.9 can be drawn to represent the external forces. The polygon can then be completed and the forces scaled off from it, the scaled values being given in Table 1.1.

Notes

(1) It is best to make a rule always to move round a joint or frame in the same direction—say, clockwise. For example, if one moves clockwise round joint *D*, the 2500 N force will be called *JG* according to Bow's notation.

(2) When drawing the force polygon for joint *D*, the force *JG* is known to be 2500 N vertically down, and it is therefore represented by a vertical line scaled to represent 2500 N and lettered *j* at the top and *g* at the bottom so that reading the letters naming the force in sequence gives the direction of the force. Still moving clockwise round joint *D*, the force in member *CD* will be called *GL*. The line *g* to *l* on the force polygon is upward and to the right, and the force at end *D* of member *CD* is therefore upward and to the right; a force arrow in this sense can therefore be put at the right-hand end of *CD* in the space diagram.

(3) When a force arrow has been put at one end of a member in the space diagram, an arrow in the opposite sense can be put at the other since the member itself must be in equilibrium under equal and opposite forces. Notice that force arrows should never be drawn on the force polygon since directions on this figure are obtained by the sequence of letters.

(4) In drawing the force polygon one can only build it as a series of lines or triangles by working at joints with three or fewer unknown

values; it may be necessary to construct the polygon in a particular order.

(5) The magnitudes of forces in members are obtained by scaling the force polygon and the types of the force by referring to the arrows on the space diagram—arrows pointing towards the ends of a member indicate that it is in compression, arrows pointing towards the middle indicate that it is in tension.

[Reaction A: 3650 N at 175° to AE. Reaction B: 4160 N along line BC.]

TABLE 1.1

Member	Force (N)	Type
AC	600	T
BC	4160	C
CD	1835	C
DE	2240	T
EC	2240	C
AE	3165	T

The forces have been obtained by scaling Fig. 1.9.

1.9. A weight of 10 N is suspended from three cords each of length 2 m fastened to three hooks set in a horizontal ceiling such that the joint between the cords P is 1 m below the level of the ceiling, and the cords are 120° to each other in plan view. Find the tension in each cord.

Solution

Take the origin of coordinate axes at point P; let the positive directions be as shown in Fig. 1.10. Assume all unknown forces are tensile.

Consider the equilibrium of point P.

$$\Sigma F_x = 0 \quad \text{positive to the right.}$$

$\therefore \; -F_{AP} \cos (\text{angle between } AP \text{ and } x \text{ axis}),$
$+F_{BP} \cos (\text{angle between } BP \text{ and } x \text{ axis}),$
$+F_{PC} \cos (\text{angle between } PC \text{ and } x \text{ axis}) = 0.$

STATICS

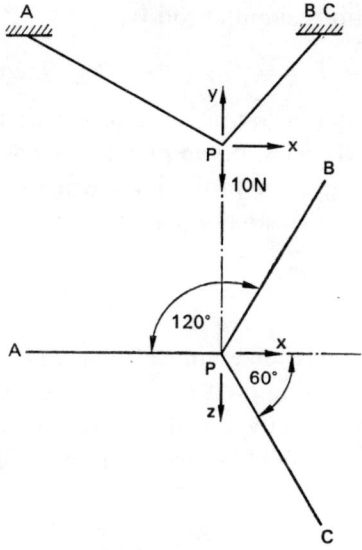

Fig. 1.10.

Inspection of the cosines indicates that each is equal to:

$$\frac{\text{Projection of the cord on the } x \text{ axis}}{\text{True length of cord}}.$$

$$\therefore \quad \frac{-\sqrt{(3)}}{2} F_{AP} + \frac{\sqrt{(3)}/2}{2} F_{BP} + \frac{\sqrt{(3)}/2}{2} F_{PC} = 0,$$

$$-2F_{AP} + F_{BP} + F_{PC} = 0. \quad (1)$$

$\Sigma F_z = 0$ positive downwards,

$$\frac{\frac{3}{2}}{2} F_{PC} - \frac{\frac{3}{2}}{2} F_{PB} = 0,$$

$$F_{PC} = F_{PB}. \quad (2)$$

$\Sigma F_y = 0$ positive upwards,

$$\tfrac{1}{2} F_{AP} + \tfrac{1}{2} F_{BP} + \tfrac{1}{2} F_{CP} - 10 = 0,$$

$$F_{AP} + F_{BP} + F_{CP} = 20. \quad (3)$$

Solving these equations simultaneously,

$$F_{AP} = F_{BP} = F_{CP} = \tfrac{20}{3} = 6\tfrac{2}{3} \text{ N tensile.}$$

(*Note:* It may seem easier in some problems to calculate angles between forces and the axes, but in practice one usually knows or can measure the true lengths or projections, while the angles cannot be measured easily with the same degree of accuracy.)

Problems

1.10. Draw a free-body diagram showing and naming the major forces and moments acting on a motor-cycle when it is being ridden at a constant high velocity. (N.B.—Show distributed forces as a resultant force acting through an assumed centre of pressure or centre of gravity.)

1.11. Figure 1.11 shows a very simple form of support for an aircraft jet engine suspended beneath the wing. Draw a free-body diagram showing the forces on the engine when it is propelling the aircraft in steady, level flight. Assume the centre line of the engine is horizontal, that the centres of thrust, pressure, and gravity all lie on the centre line, and that all the joints are frictionless pins.

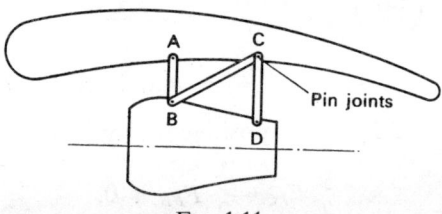

Fig. 1.11.

1.12. A vertical post *CB* is rigidly set in concrete and a horizontal arm *AB* is rigidly bolted to the upright at *B* so that the free end *A* is 2 m vertically above and 1 m to the right of *C*. Assuming the post and bar are weightless, draw free-body diagrams of the two parts showing all forces and moments acting on them when a force of 10 kN acts vertically down at A.

[The answer is shown in Fig. 1.12.]

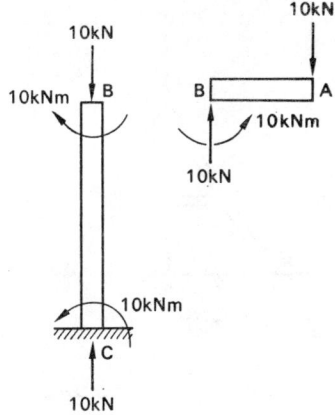

Fig. 1.12.

1.13. A horizontal cantilever BC of length L is rigidly built in to a wall at C. At the end B a right-angle bracket is firmly welded to the cantilever and a load W is applied as shown in Fig. 1.13. Draw free-body diagrams (a) of the bracket, and (b) of the main cantilever, showing all forces and moments acting on them.

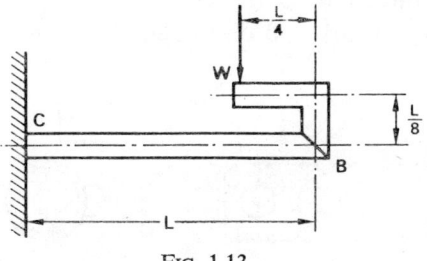

Fig. 1.13.

1.14. A plane vertical frame rests on a smooth horizontal floor as shown in Fig. 1.14. The two side members AB and AC are each of length $2L$ and are pinned together at A. The cross-bar DE has a length

L and is pinned to the mid-points of the side members. Draw free-body diagrams of all three members showing the forces due to a vertically downward load W acting on the pin at A.

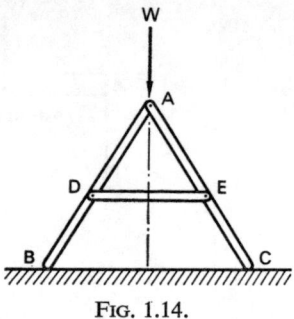

FIG. 1.14.

1.15. A cylindrical bar protrudes horizontally from a vertical wall. At a distance L from the wall is a horizontal cross-handle extending a distance d on either side of the bar. Find the reactive forces and moments relative to the x, y, and z axes shown in Fig. 1.15 when:
  (a) equal loads P act vertically down at each end of the cross-handle;
  (b) equal loads P act vertically at each end but the one on the left is downward and on the right upward when looking in the positive x direction;
  (c) only the downward load on the left is acting.

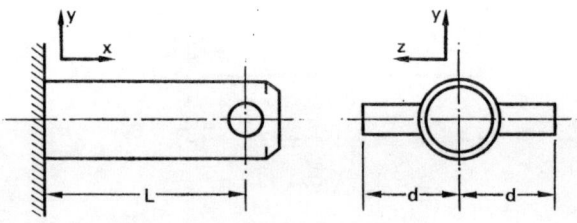

3rd Angle projection

FIG. 1.15.

[(a) $M_x = M_y = 0$, $M_z = +2PL$; $R_x = R_z = 0$, $R_y = +2P$.
(b) $M_x = +2Pd$, $M_y = M_z = 0$; $R_x = R_y = R_z = 0$.
(c) $M_x = +Pd$, $M_y = 0$, $M_z = +PL$; $R_x = R_z = 0$, $R_y = +P$.]

1.16. An aircraft undercarriage leg and wheel have the configuration shown. During landing, the wheel is assumed to exert forces of 7 kN vertically and 3.5 kN horizontally on the leg through the centre of the wheel bearing. What reactions must the mounting exert on the leg? Give the answer relative to the axes shown in Fig. 1.16.

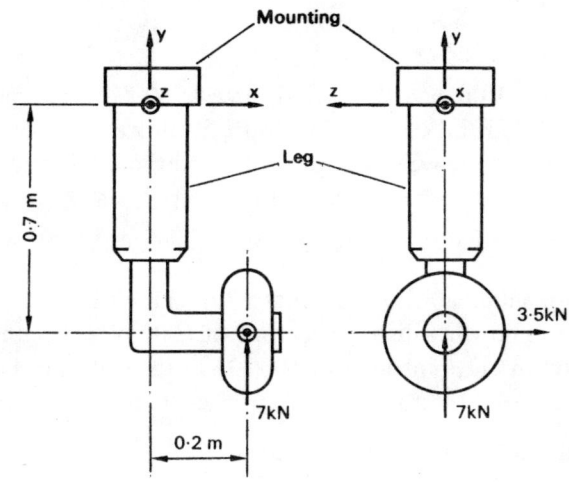

3rd Angle projection

FIG. 1.16.

[$R_x = 0$, $R_y = -7$ kN, $R_z = 3.5$ kN.
$M_x = -2450$ N m, $M_y = 700$ N m, $M_z = -1400$ N m.]

1.17. A sign board is mounted at the top of a vertical pole as shown in Fig. 1.17. The weight of the board is 100 N and the centre of gravity

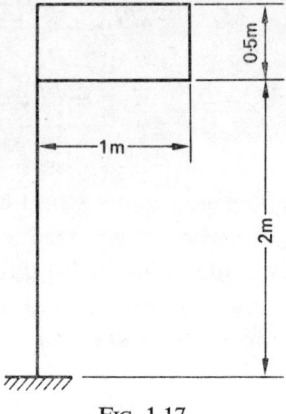

Fig. 1.17.

is at the centre of the board. A wind force of 50 N acts perpendicular to the board, and the centre of pressure is also at the centre of the board. Neglecting the weight of the pole, find the total resultant force and total resultant moment exerted by the pole on its foundations.

[Resultant force 112 N. Resultant moment 125·8 N m.]

1.18. A horizontal cantilever of length 3 m is built in to a rigid wall; it supports a uniformly distributed load of 6000 N/m length over the whole length and a point load of 5000 N at the free end, both loads acting downwards. Find the supporting reactions at the wall.

[23 kN. 42 kN m.]

1.19. A three-hinged arch is symmetrical about the vertical centre line. The span is 36 m and the rise to the centre hinge is 6 m. The arch carries point loads vertically downwards as follows: at 3 m from the left-hand support, 50 kN; at 9 m from the left-hand support, 90 kN; and at 15 m from the left-hand support, 80 kN. Find the vertical and horizontal components of the reactions at each of the three hinges and show them on diagrams of the separate halves of the arch.

[The answer is shown in Fig. 1.18.]

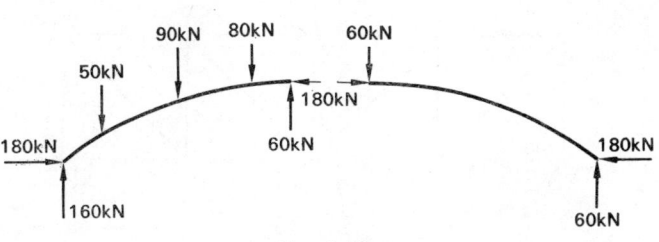

FIG. 1.18.

1.20. State the conditions required in order that the force in a member of a plane frame shall be along the length of the member itself. Specify and calculate the forces and/or couples acting in the members of the frame shown in Fig. 1.19. The members *AB*, *BC*, and *CD* in the figure

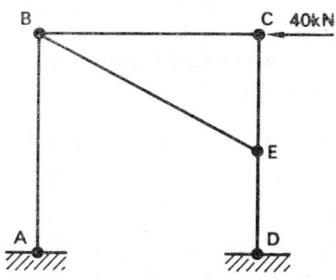

FIG. 1.19.

are all straight and 3 m long, pin-jointed at each end. *CD* is one continuous member drilled at its mid-point *E* to accommodate a pin connection for *BE*. A load of 40 kN acts horizontally at *C*, and the frame is pinned to the ground at *A* and *D*.

[London Univ., B.Sc. 1]

[*AB* 40 kN compression. *BC* 80 kN compression. *BE* 89·5 kN tension. *EC* couple 60 kN m. *ED* couple 60 kN m plus 40 kN tension.]

1.21. The girder shown in Fig. 1.20 has four square bays and all joints are pinned. Both supports are at the same horizontal level,

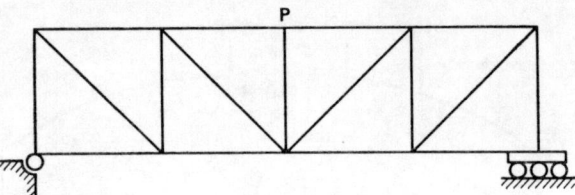

Fig. 1.20.

the left-hand is a pin joint and the right a frictionless roller. Find the type of force (magnitude is not required) in each member under the following three loading conditions:

(a) A load of 8 units is applied vertically downwards at *P*.
(b) A load of 4 units vertically down and 4 units horizontally to the left is applied at *P*.
(c) A load of 4 units vertically down and 4 units horizontally to the right is applied at *P*.

[(a) Tension in all diagonals and two centre lower horizontals; compression in verticals and four upper horizontals; no force in two outer lower horizontals.
(b) Compression in all upper horizontals and all verticals; tension in diagonals; lower horizontals, compression in two on left, tension in centre right, no force in outer right.

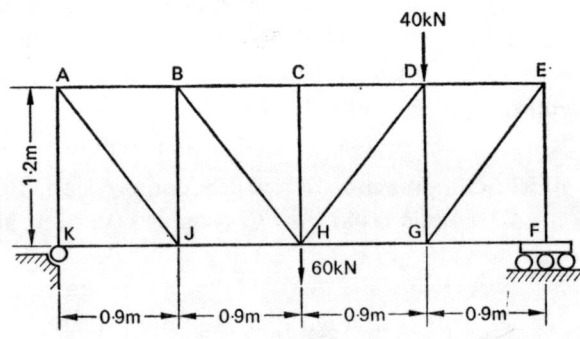

Fig. 1.21.

(c) Compression in all upper horizontals and all verticals, tension in all diagonals and lower horizontals except right outer which has no force.]

1.22. Find the magnitude and type of the force in each member of the pin-jointed V-girder shown in Fig. 1.21.

[The answer is contained in Table 1.2.]

TABLE 1.2

Member	Force (kN)	Type
AK, BJ	40	C
JK, CH, FG	0	—
AJ, BH	50	T
AB, JH, GH	30	T
BC, CD, EF, GD	60	C
EG	75	T
ED	45	C
HD	25	T

1.23. The pin-jointed girder shown in Fig. 1.22 has a span of 12 m, a height of 3 m, and every panel is a square. Find the forces in mem-

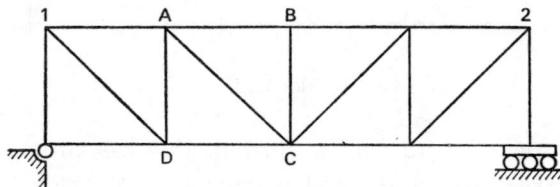

Fig. 1.22.

bers *AB*, *AC*, and *CD* when the girder is carrying vertical loads of 10 kN at joints 1 and 2 and vertical loads of 20 kN at each of the other three top joints.

[The answer is given in Table 1.3.]

TABLE 1.3

Member	Force (kN)	Type
AB	44·14	C
AC	14·14	T
DC	30	T

1.24. The three-pinned arch shown in Fig. 1.23 is made up of pin-jointed members. Confirm that the reactions are as follows:

[$V_A = 600$ kN. $H_A = 400$ kN. $V_H = 200$ kN. $H_H = 400$ kN.]

and the forces in the members are as shown in Table 1.4.

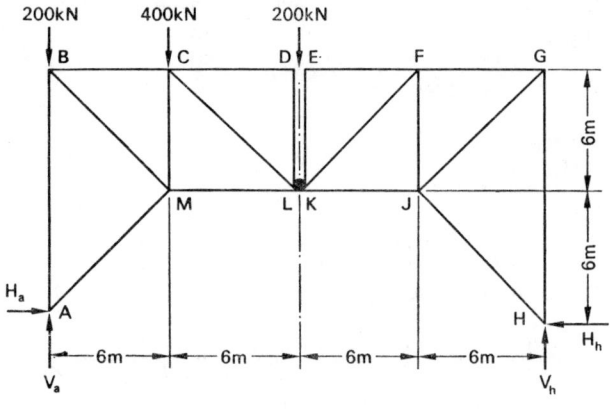

FIG. 1.23.

1.25. The Warren girder drawn in Fig. 1.24 has all its members of length L. The span is $4L$ and it carries equal loads W vertically downwards at points F, G, and H. If the maximum allowable force in any of the upper horizontal members is 500 kN, find (a) the magnitude of W, (b) the size and location of the greatest force in the lower horizontal members, and (c) the size and location of the maximum force which occurs in a diagonal.

[London Univ., B.Sc. 1]

TABLE 1.4

Member	Force (kN)	Type
EK, EF, BM, BC, CD, CL	0	—
ML, MC	400	C
FJ, AB, DL	200	C
AM, HJ	565·6	C
KJ	600	C
KF	282·8	T
GJ	282·8	C
GH, FG	200	T

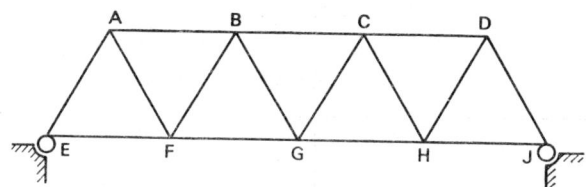

Fig. 1.24.

(*Hint:* Assume $W = 1$ and find the maximum force F_{max} in an upper horizontal. The force in this member will be proportional to the size of W, and W can be found from $W \times F_{max} = 500$.)

[$W = 216·2$ kN; 437·5 tension in *FG* and *GH*. 375 kN compression in *AE* and *DJ*. 375 kN tension in *AF* and *DH*.]

1.26. A roof-truss of dimensions and loading shown in Fig. 1.25 is pin-jointed throughout and pinned to a rigid support at *A*. The other support *B* is on a roller at the same level as *A*. Use the method of sections to find the forces in members *CE*, *CD*, and *FE*.
[The answer is given in Table 1.5.]   [London Univ., B.Sc.]

1.27. The plane pin-jointed frame shown in Fig. 1.26 carries a load of 200 kN as shown. Find, by calculation, the forces in the four members that meet at joint *B*.
[The answer is given in Table 1.6.]   [London Univ., B.Sc.]

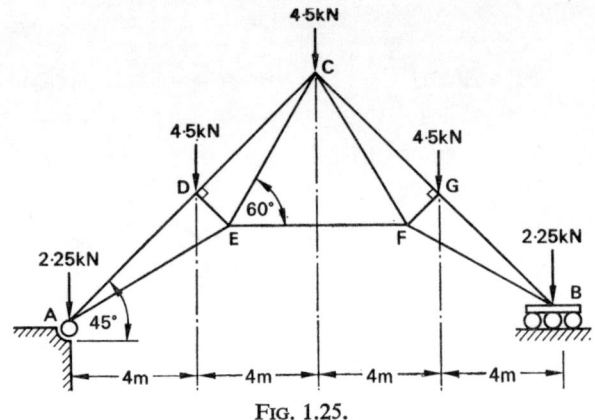

Fig. 1.25.

Table 1.5

Member	Force (kN)	Type
FE	7·3	T
CE	13·8	T
DC	20·4	C

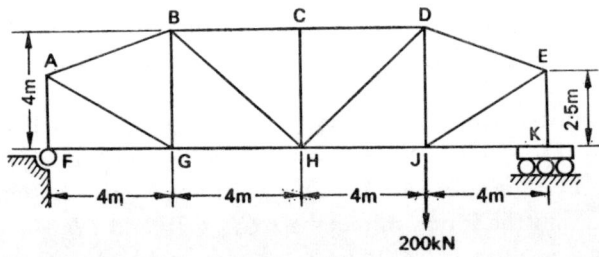

Fig. 1.26.

1.28. Figure 1.27 shows a plane pin-jointed frame which carries vertical loads of 3 units at B, C, and D and horizontal forces of 2 units at B and C. There is no horizontal component of reaction at A. Find

STATICS

TABLE 1.6

Member	Force (kN)	Type
AB	51·07	C
GB	32·07	C
HB	70·70	T
CB	100	C

either graphically or by calculation the forces (magnitude and type) in members *BC*, *BF*, *CF*, and *FG*. Lever arms of forces may be taken from a scale drawing. [London Univ.]

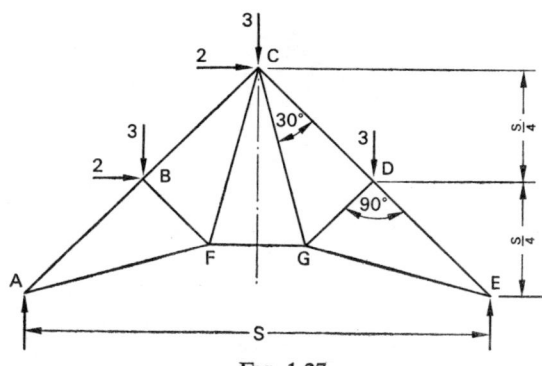

FIG. 1.27.

(*Note:* If a scale drawing is to be made to enable lever arms to be measured, one might just as well carry on with a graphical solution. Using such a solution one obtains the values in Table 1.7.)

TABLE 1.7

Member	Force (units)	Type
BC	4·4	C
BF	3·5	C
CF	3·7	T
FG	0·75	T

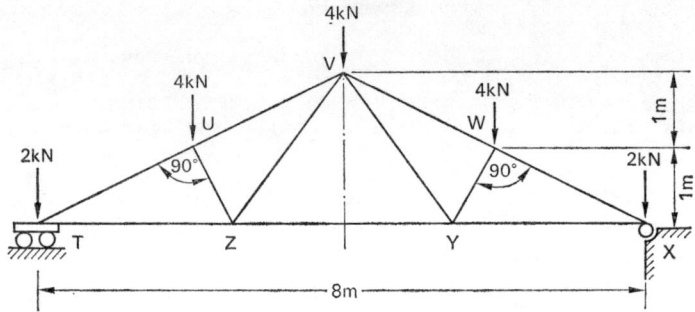

Fig. 1.28.

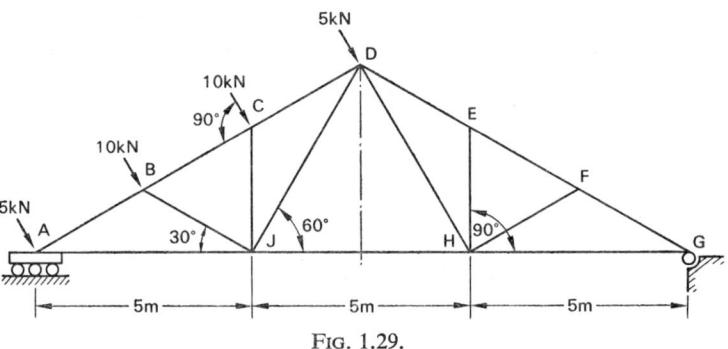

Fig. 1.29.

1.29. Find the forces in all members of the plane roof-trusses shown (a) in Fig. 1.28, and (b) in Fig. 1.29. Both trusses are pin-jointed throughout and mounted on two supports at the same horizontal level; the left-hand support is a roller and the right-hand a pin.
[The answer is given in Table 1.8.]

1.30. The pin-jointed frame shown in Fig. 1.30 is symmetrical about the vertical centre line, supported on a hinge at A and on rollers at D. The span is 18 m, all bottom horizontal members are of equal length, and the posts are of lengths $GC = 6$ m, $FH = 1.5$ m, $EB = 3$ m. Find the forces in all members when vertical loads of 15 kN act downwards at joints E, F, and G.
[The answer is given in Table 1.9.] [London Univ., B.Sc.]

STATICS

TABLE 1.8

	Member	Force (kN)	Type
(a)	TU, WX	13·45	C
	UV, VW	11·75	C
	TZ, YX	12·00	T
	UZ, WY	3·60	C
	ZY	8·00	T
	VZ, VY	7·20	C
(b)	AB, CD	26·0	C
	BC	20·0	C
	AJ	20·0	T
	DE, EF, FG	17·6	C
	BJ, CJ	11·8	C
	JD	20·0	T
	HG, JH, DH, EH, FH	0	—

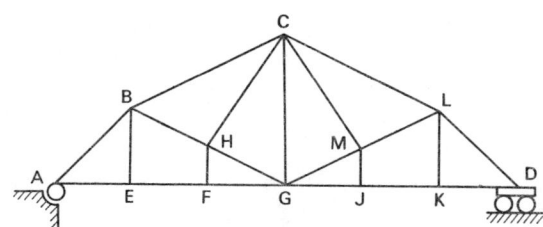

FIG. 1.30.

1.31. The "Fink" roof-truss shown in Fig. 1.31 has a span of 16 m. The apex is 4 m above the level of the supports and the horizontal member 11–12 is 0·4 m above the level of the supports. The loads are perpendicular to the rafters and members 1–2, 2–3, 3–4, 4–5 are all the same length. Find the forces in all members.

Solution

It will be found that a force diagram can be drawn for the loads and reactions and for joints 1, 2, 13, 8, 9, and 10 but then the next joints

TABLE 1.9

Member	Force (kN)	Type
KD, BE, FH, CG, GJ, JK	15	T
AE, EF, FG	30	T
AB	42·4	C
CB	33·4	C
CH	13	T
HG	8	C
GM, ML	9	T
DL	21·2	C
CL	25	C
BH, CM, JM, LK	0	—

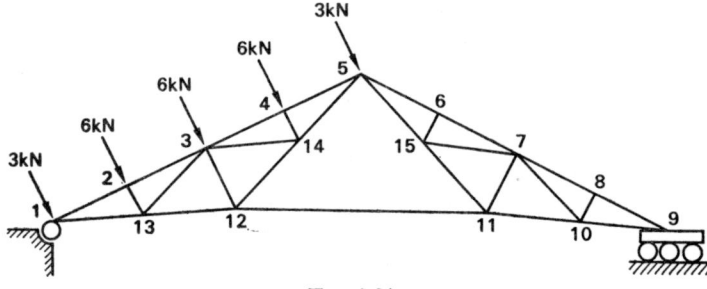

FIG. 1.31.

3, 12, 7, and 11 all have three unknown forces and the diagram cannot be completed. It is necessary to carry out an imaginary modification of the structure by replacing members 3–14 and 4–14 by a single member 12–4; it will be found that this could be done without making the structure collapse. The polygon is then continued and can be completed treating 4–12 as a real member. Once the rest of the diagram is finished, the substitution is reversed and the vectors for members 3–14 and 4–14 can be located within the diagram.
[Answer is given in Table 1.10.]

1.32. The three legs of a tripod are each 4 m long and are fixed to the ground at the vertices of an equilateral triangle of 3 m side. Find the

TABLE 1.10

Member	Force (kN)	Type
1–2, 2–3, 3–4, 4–5	34	C
5–6, 6–7, 7–8, 8–9	18	C
6–15, 7–15, 7–11, 7–10, 8–10	0	—
1–13	40	T
13–12	32	T
12–11	15	T
11–10, 9–10	16	T
2–13, 4–14	6	C
3–12	12	C

force in each leg due to a weight of 1 kN hanging from the apex of the tripod.

[370 N compression.]

1.33. Four hooks are fixed at the corners of a square of side 2 m marked on a horizontal ceiling. A cord is attached to each hook and all four are joined together at a point 1 m vertically below the centre of the square from which point a mass weighing 10 N is suspended. Calculate the tension in each cord.

[6·66 N.]

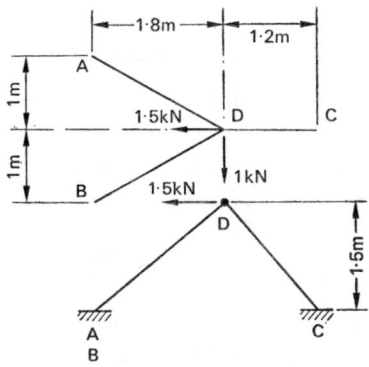

3rd Angle Projection

FIG. 1.32.

1.34. Figure 1.32 shows two views in third-angle projection of a tripod supporting two horizontal loads. Assuming the members are pinned to a horizontal floor and to each other, find the force in each member. [London Univ.]
[The answer is given in Table 1.11.]

1.35. A pair of shear legs is constructed from two members each 13 m long. The bases are pinned to anchorage points set 6 m apart in horizontal ground. The legs are pinned together at the end and inclined so that the joint is vertically above a point 3·5 m to the right of the mid-point of the line joining the anchorages. A wire stay is fitted to the apex and anchored to the ground 18·5 m to the left of

TABLE 1.11

Member	Force (kN)	Type
AD	0·635	T
BD	1·903	C
DC	0·956	T

the mid-point of the line joining the leg anchorages. Find the forces in each leg and in the stay when a weight of 300 kN is suspended from the apex.

[190·5 kN in each leg. 117 kN in the stay.]

1.36. Two light bars are pinned together at one end and to two points on a vertical wall at the other. They are retained in position by a cable also attached to the wall at the point shown in Fig. 1.33. Find the force in the bars and cable when a vertical load of 4500 N acts down at the joint.

[1590 N in the legs. 3895 N in the cable.]

1.37. A tripod of dimensions shown in Fig. 1.34 is pinned to a vertical wall at points A, B, and C and there is a pin joint at O where a force of 100 kN acts vertically downwards. Find the forces in all members.
[London Univ., B.Sc.]

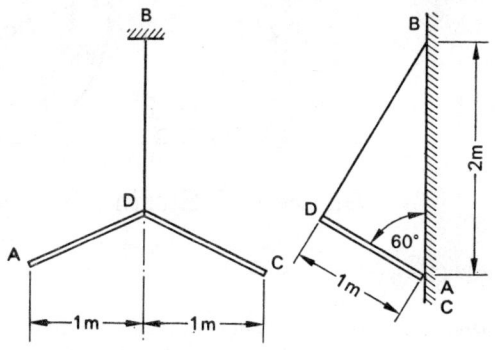

3rd Angle Projection[N]

Fig. 1.33.

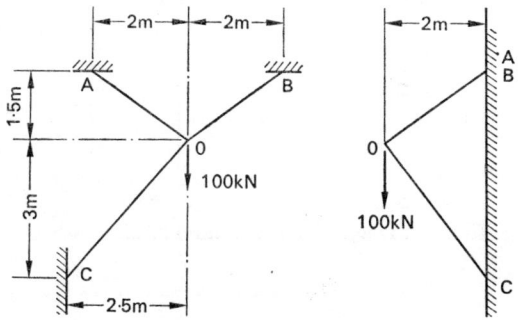

3rd Angle Projection

Fig. 1.34.

Table 1.12

Member	Force (kN)	Type
OB	8·7	C
OA	80·2	T
OC	97·6	C

CHAPTER 2

Stress and Strain

Definitions and Theory

(a) *Direct stress*. The direct stress f set up by a tensile or compressive force F acting perpendicular to the cross-sectional area A of an elastic material (Fig. 2.1) is defined as $f = F/A$ with the sign convention that tensile forces and stresses are positive and compressive forces and stresses are negative.

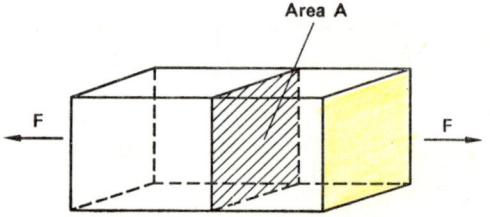

FIG. 2.1. Diagram for definition of direct stress.

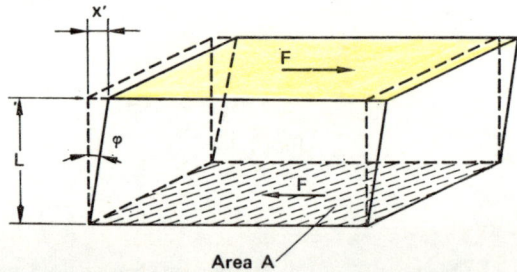

FIG. 2.2. Diagram for definition of shear stress and strain.

(b) *Shear stress*. Parallel forces F acting in opposite directions on a material as in Fig. 2.2 produce shear stresses q, where $q = F/A$, where A is not the cross-sectional area but the area being sheared.

(c) *Strain.* This is the measure of deformation of a material carrying stress. Direct strains are caused by direct stresses and have the symbol e, where

$$e = \frac{\text{change in length}}{\text{original length}} = \frac{x}{L};$$

compressive strains negative, tensile positive. Shear strain symbol φ is defined with reference to Fig. 2.2 as $\varphi = x'/L$ but since strains in an elastic material are usually very small, the shear strain is approximately equal to the angle φ measured in radians. Strain is a dimensionless number and therefore has no units.

(d) *Elastic moduli.* Within the range of stresses and strains used in many engineering applications it is found that the ratio of stress to strain is a constant. The constant is called the elastic modulus and is named for uniaxial direct tension or compression, shear, and for hydrostatic pressure as:

(i) Uniaxial direct stress: Young's modulus or modulus of elasticity (E).
(ii) Shear: Modulus of rigidity (G).
(iii) Hydrostatic pressure (equal compressive stress on all surfaces of a solid body).
Bulk modulus (K) defined as $K = p/(\Delta v/v)$ where p is the pressure, Δv is the change in volume, v is the original volume, and $\Delta v/v$ is called volumetric strain.

(e) *Poisson's ratio.* If a body is being strained elastically by uniaxial direct stress (Fig. 2.1), it is found that not only is there strain in the direction of the stress (longitudinal strain) but also lateral strain in perpendicular directions. The ratio

$$\frac{\text{Lateral strain}}{\text{Longitudinal strain}}$$

is known as Poisson's ratio (ν).

Since a tensile longitudinal strain gives rise to a compressive lateral strain, the sign of the ratio is negative, but it is usual to keep this in mind and to quote only the absolute value of v.

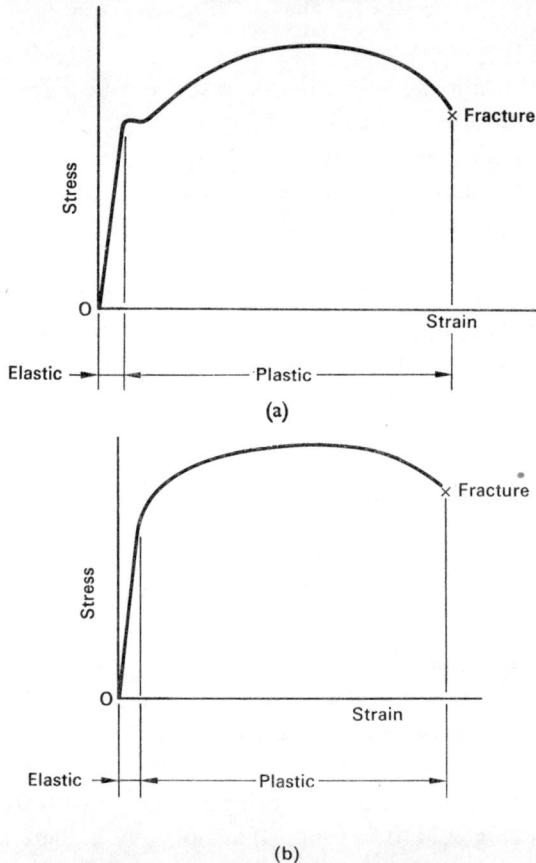

FIG. 2.3. Typical stress/strain diagrams. (a) For a material with a yield point. (b) For a material without a definite yield point.

(f) *The tensile test.* If a specimen of material is subjected to uni-axial tension under carefully standardized conditions and the load and extension measured, graphs of stress/strain can be plotted from which important design parameters for the material can be obtained. Figure 2.3 shows the form of typical graphs for metals. Initially there is a linear portion where stress is proportional to strain, but at some point—known as the limit of proportionality—a departure from linearity can be observed. In the same region it is found that if the stress is returned to zero there will be some permanent strain remaining; the elastic limit is the stress at which this "permanent set" first occurs. Just beyond the elastic limit some materials exhibit a definite "yield" at which there is a sudden increase in strain with no corresponding increase in stress. Other materials have no such yield, but it is common to speak of a yield stress defined as that stress at which plastic deformation begins. It is common, too, to use the proportional limit, elastic limit, and yield stress as alternatives for one another in design calculations: since they are usually close to one another, this is quite acceptable.

Other values obtained from a tensile test are:

(i) *Proof stress*—that stress which when applied and removed causes a definite percentage permanent set (usually 0·1 per cent).
(ii) *Ultimate stress*—the greatest load that can be applied divided by the original cross-sectional area.
(iii) *Percentage elongation*—the percentage increase in the gauge length when the specimen is loaded to fracture.
(iv) *Percentage reduction in area*—the percentage decrease in the cross-sectional area at the point of fracture compared with the original area.

(g) *Stresses in thin cylinders* (a thin cylinder is one where the wall thickness t is very much less than the diameter). Hoop stress f_H due to internal pressure $p = pr/t$, where r is the internal radius. If the ends of the cylinder are closed, there will be a longitudinal stress $f_L = pr/2t$. In a thin cylinder the stress in the radial direction is so small compared with f_L and f_H that it can be assumed to be zero.

(h) *Thermal stresses*. Most materials undergo dimensional changes as their temperature changes. If the dimensional change is restricted, thermal stresses will be set up in the material. Problems are best solved by imagining that the material is free and does change its dimension but is then forced to return to the known size by external loads. The stress which such loads would cause is then equal to the thermal stress, e.g. if a bar of length L, coefficient of linear expansion α, has its length rigidly restrained but undergoes a temperature rise t, its length would become $L(1+\alpha t)$.

To return it to its original length it must be forced back a distance αLt. The strain on it is

$$\frac{\alpha Lt}{L(1+\alpha t)}.$$

Since αt is very small compared with unity, strain $\doteqdot (\alpha Lt)/L = \alpha t$.
But stress/strain $= E$.
Therefore thermal stress $= E\alpha t$.

(i) *Elastic strain energy.* When a body is strained within the elastic limit, the external work done is stored within the material as strain energy (also called resilience). It can be shown that a direct stress f acting uniformly throughout a volume of material gives rise to strain energy of $(f^2/2E)\times$volume.

The unit of strain energy—as of all other forms of energy—is the Joule (1 J = 1 N m).

The proof resilience of a material is the strain energy it will store (per unit volume) when the stress is at the elastic limit.

(j) *Impact loading.* If a weight W falls vertically through a distance h on to a prismatic bar of uniform cross-section A and length L, which is deformed elastically in arresting the fall, then the maximum stress set up in the material is given by

$$f = \frac{W}{A} + \sqrt{\left[\left(\frac{W}{A}\right)^2 + \frac{2WhE}{AL}\right]},$$

a relation obtained by equating the potential energy, relative to a datum at the lowest point reached when the bar deflects, of the weight before dropping to the strain energy $(f^2AL)/2E$ stored in the bar at maximum deflection.

Worked Examples

2.1. An aluminium rod 25 mm diameter and 1 m long carries an axial tensile load of 20 kN. Find (a) the tensile stress, (b) the increase in length, (c) the decrease in diameter, and (d) the change in volume due to the load. ($E = 69$ GN/m². $v = \frac{1}{3}$.)

Solution

(a) Stress = load/area = $20 \times 10^3 / \frac{1}{4}\pi \; 0{\cdot}025^2$
$$= 40{\cdot}7 \times 10^6 \text{ N/m}^2.$$

(b) Strain = change in length/original length,
E = stress/strain.

$\therefore$ change in length = $\dfrac{\text{stress}}{E} \times$ original length

$$= \frac{40{\cdot}7 \times 10^6}{69 \times 10^9} \times 1$$

$$= 5{\cdot}9 \times 10^{-4} \text{ m}$$

$$= 0{\cdot}59 \text{ mm}.$$

(c) $v = \dfrac{\text{lateral strain}}{\text{longitudinal strain}}$.

$\therefore$ lateral strain = $\dfrac{1}{3} \times \dfrac{5{\cdot}9 \times 10^{-4}}{1}$.

$$\frac{\text{Change in diameter}}{0{\cdot}025} = \frac{5{\cdot}9 \times 10^{-4}}{3}.$$

Change in diameter = $4{\cdot}92 \times 10^{-6}$ m
$$= 0{\cdot}0049 \text{ mm}.$$

(d) Suppose the strain on a diameter = e_D and the longitudinal strain = e_L.

Length under load = $L(1+e_L)$.

Area of cross-section = $\dfrac{\pi D^2 (1+e_D)^2}{4}$.

∴ new volume $= L(1+e_L)\dfrac{\pi D^2}{4}(1+e_D)^2$.

Change in volume $= \dfrac{\pi D^2 L}{4}(1+e_L)(1+e_D)^2 - \dfrac{\pi D^2 L}{4}$.

When the brackets are multiplied together there will be terms which are products and powers of strains. Since strains are small numbers all such products and powers can be discarded and

$$\text{change in volume} = \dfrac{\pi D^2 L}{4}(e_L + 2e_D)$$

$$= \dfrac{\pi \times 0.025^2 \times 1}{4}(5.9 \times 10^{-4} - 2 \times 1.97 \times 10^{-4})$$

$$= \dfrac{\pi \times 0.025^2}{4}(1.96 \times 10^{-4})$$

$$= 0.097 \times 10^{-6} \text{ m}^3.$$

2.2. A rigid bar 0·6 m long is supported in the horizontal position by two vertical tie rods 10 m long and 12 mm diameter attached at its ends. The left-hand rod is of wrought iron ($E = 200$ GN/m²) and the right-hand rod of brass ($E = 100$ GN/m²). At what distance along the bar from the left-hand end must a 5 kN load be placed if the bar is to remain horizontal? What are the stresses in the rods?

Solution

The general method of attack on problems involving loads supported by members of two different materials is to form two equations.

(1) Total load = load in material A + load in material B.
(2) Strain in material A = $C \times$ (strain in material B), where C is a number obtained from consideration of the form of the problem.

STRESS AND STRAIN

In this problem suppose f_I and f_B are the stresses in iron and brass respectively and A_I and A_B the cross-sectional areas.

$$\text{Total load} = f_I A_I + f_B A_B.$$
$$\therefore\ 5000 = f_I A_I + f_B A_B. \tag{1}$$

Since the bar is horizontal both before and after loading, the extension of each bar is the same, and since they are both 10 m long,

$$\text{strain in iron rod } (e_I) = \text{strain in brass rod } (e_B). \tag{2}$$

But
$$\frac{f}{e} = E.$$

Therefore substituting in eqn. (2),

$$\frac{f_I}{E_I} = \frac{f_B}{E_B},$$

$$f_I = f_B \frac{E_I}{E_B},$$

$$f_I = 2 f_B.$$

Substituting this in eqn. (1),

$$5000 = f_B(2A_I + A_B).$$

But
$$A_I = A_B = 113 \cdot 1 \times 10^{-6}\ \text{m}^2,$$

$$f_B = \frac{5000}{(339 \cdot 3 \times 10^{-6})}\ \text{N/m}^2$$
$$= 14 \cdot 75\ \text{MN/m}^2,$$
$$f_I = 29 \cdot 5\ \text{MN/m}^2$$

To find the position of the load, suppose it is at d from the left-hand end and consider the equilibrium about that end.

Σ (moments about left-hand end) $= 0$ clockwise positive.

$$5000d - 14 \cdot 75 \times 10^6 \times 113 \cdot 1 \times 10^{-6} \times 0 \cdot 6 = 0$$

$$d = \frac{14 \cdot 75 \times 113 \cdot 1 \times 0 \cdot 6}{5000},$$

$$d = 0 \cdot 2\ \text{m}.$$

2.3. A cylindrical mild steel bar of 0·02 m diameter and length 0·3 m is placed inside a tube of the same length and of internal diameter 0·02 m, external diameter 0·026 m. The combination is then subjected to an axial thrust of 50 kN. Find (a) the stress in the tube and in the rod, and (b) the amount the rod shortens. (E_{steel} = 206·8 GN/m². $E_{tube\ material}$ = 80 GN/m².) [London Univ.]

Solution

Let stress, strain, and cross-sectional area of tube be f_T, e_T, and A_T respectively and let stress, strain, and area of bar be f_B, e_B, and A_B.

Total load = load carried by bar + load carried by tube,

$$50 \times 10^3 = f_B A_B + f_T A_T. \tag{1}$$

Both bar and tube shorten the same amount and have the same original length.

$$\therefore e_T = e_B.$$

But

$$e = \frac{f}{E}.$$

$$\therefore f_T = \frac{E_T}{E_B} f_B. \tag{2}$$

$$A_B = \frac{\pi \times 0.02^2}{4} = 314.2 \times 10^{-6}\ \text{m}^2, \quad A_T = \frac{\pi(0.026^2 - 0.02^2)}{4}$$

$$= 216.7 \times 10^{-6}\ \text{m}^2,$$

while

$$\frac{E_T}{E_B} = \frac{80}{206.8} = 0.387.$$

$$\therefore f_T = 0.387 f_B, \text{ and substituting in eqn. (1) gives}$$

$$50 \times 10^3 = 314.2 \times 10^{-6} f_B + 216.7 \times 10^{-6} \times 0.387 f_B,$$

$$f_B = 125.6 \times 10^6\ \text{N/m}^2,$$

$$f_T = 48.6 \times 10^6\ \text{N/m}^2,$$

$$e_T = \frac{f_T}{E_T} = \frac{48.6 \times 10^6}{80 \times 10^9}.$$

$$\therefore \text{shortening} = 0.3\ e_T = 0.182\ \text{mm}.$$

2.4. A bar 0·6 m long is made up of a steel rod 0·2 m long, 30 mm diameter fixed concentrically at one end to a rod of copper 0·4 m long. Find the diameter of the copper rod in order to make the extensions of each material the same under an axial tensile load of 20 kN. What will then be the stresses in each material? ($E_{steel} = 200$ GN/m². $E_{Cu} = 110$ GN/m².)

Solution

Referring to Fig. 2.4,

$$\text{area of steel} = \frac{\pi}{4} \times (0.03)^2 = 7.069 \times 10^{-4} \text{ m}^2.$$

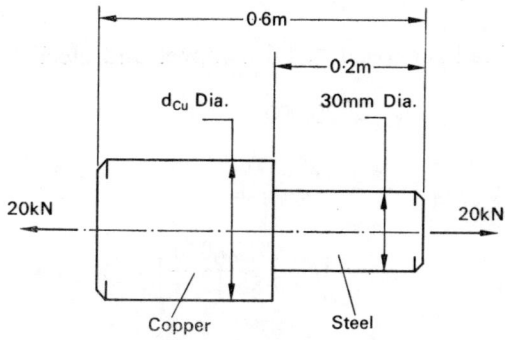

Fig. 2.4.

Using suffix s for steel and Cu for copper, L for original length, and δ for extension,

$$e_s = \frac{f_s}{E_s}.$$

But
$$e_s = \frac{\delta_s}{L_s}.$$

$$\therefore \delta_s = \frac{L_s f_s}{E_s}.$$

Similarly,
$$\delta_{Cu} = \frac{L_{Cu} f_{Cu}}{E_{Cu}}.$$

But the extensions are to be equal.

$$\therefore \frac{L_s f_s}{E_s} = \frac{L_{Cu} f_{Cu}}{E_{Cu}},$$

$$f_{Cu} = \frac{E_{Cu}}{E_s} \frac{L_s}{L_{Cu}} f_s.$$

But
$$f_s = \frac{20 \times 10^3}{7 \cdot 069 \times 10^{-4}} = 28 \cdot 3 \times 10^6 \text{ N/m}^2,$$

$$f_{Cu} = \frac{110 \times 10^9}{200 \times 10^9} \times \frac{0 \cdot 2}{0 \cdot 4} \times 28 \cdot 3 \times 10^6,$$

$$f_{Cu} = 7 \cdot 8 \times 10^6 \text{ N/m}^2.$$

The whole load is carried by both copper and steel.

$$\therefore f_{Cu} A_{Cu} = 20{,}000,$$

$$\frac{\pi}{4} d_{Cu}^2 = \frac{20{,}000}{7 \cdot 8 \times 10^6},$$

$$d_{Cu} = \sqrt{\left(\frac{4 \times 20{,}000}{7 \cdot 8 \times 10^6 \pi}\right)} = 57 \text{ mm}.$$

2.5. A block of brass 75 mm square and 40 mm long at 15°C is placed between two perfectly rigid stops 40 mm apart. (a) Calculate the force exerted by the brass on each stop when the temperature of the bar is raised to 120°C, and (b) to what temperature could the block be heated if the stress in it is not to exceed 220 MN/m²? ($E = 100$ GN/m². $\alpha = 18 \cdot 5 \times 10^{-6}$ per °C.)

Solution

(a) Change in temperature $= 120 - 15 = 105$°C. If unrestrained the bar would expand by

$$L \alpha t = 0 \cdot 04 \times 18 \cdot 5 \times 10^{-6} \times 105 \text{ m}.$$

Strain is approximately $\dfrac{L\alpha t}{L} = 18 \cdot 5 \times 105 \times 10^{-6}$.

$$\text{Strain} = \dfrac{\text{stress}}{E}.$$

$\therefore$ stress $= 100 \times 10^9 \times 18 \cdot 5 \times 105 \times 10^{-6}$
$= 194 \times 10^6$ N/m².

Force $=$ stress $\times$ area $= 194 \times 10^6 \times (0 \cdot 075)^2$
$= 1 \cdot 09$ MN.

(b) $\quad f = E\alpha t.$

$\therefore \; t = \dfrac{f}{E\alpha}$

$= \dfrac{220 \times 10^6}{100 \times 10^9 \times 18 \cdot 5 \times 10^{-6}}$

$= \dfrac{2 \cdot 2}{1 \cdot 85} \times 10^2$

$= 119°.$

t is the temperature change so the block could be heated to 134°C.

2.6. The mild-steel bar shown in Fig. 2.5 is rigidly mounted in a vertical position. A weight of 1000 N is allowed to fall on to the

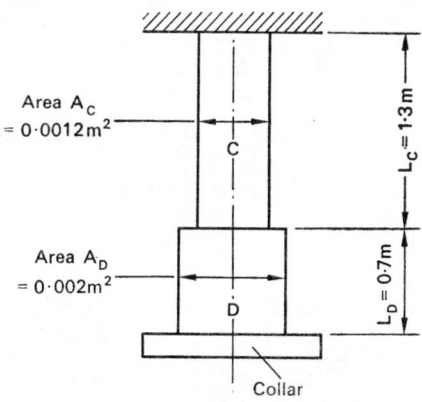

Fig. 2.5.

rigid collar at the bottom of the bar. Through what height can the weight be allowed to fall if the maximum stress in the bar is limited to 50 MN/m²? ($E = 200$ GN/m².)

Solution

The potential energy of the weight relative to a datum taken at the lowest point to which it falls is equated to the strain energy stored in the bar.

Suppose section C of the bar extends a distance δ_C and section D a distance δ_D and the stresses at the instant of these deflections are f_C and f_D.

Relative to its lowest position the weight loses potential energy of $W(h+\delta_C+\delta_D)$ joules.

The strain energy stored in the bar

$$= \frac{f_C^2}{2E} A_C L_C + \frac{f_D^2}{2E} A_D L_D ,$$

$$W(h+\delta_C+\delta_D) = \frac{1}{2E} (f_C^2 A_C L_C + f_D^2 A_D L_D),$$

$$E_C = \frac{f_C}{e_C} = \frac{f_C L_C}{\delta_C}.$$

$$\therefore \ \delta_C = \frac{f_C L_C}{E_C} \quad \text{and} \quad \delta_D = \frac{f_D L_D}{E_D}.$$

Now for equilibrium at the cross-section where the diameter changes,

$$f_C A_C = f_D A_D ,$$

$$f_C = f_D \frac{A_D}{A_C}.$$

Thus $f_C > f_D$ and $f_C = 50$ MN/m²,

$$f_D = 50 \times \frac{0 \cdot 0012}{0 \cdot 002} \times 10^6$$

$$= 30 \text{ MN/m}^2 .$$

Substituting the known values into the equation and solving for h gives

$$h = 12 \cdot 5 \text{ mm.}$$

2.7. A cylindrical steel tube with closed ends is 1000 mm long, 75 mm internal diameter, and has a wall thickness of 1·5 mm. The tube is subjected to an internal pressure p of 2750 kN/m² above atmospheric.

If $E = 200$ GN/m² and $\nu = 0 \cdot 25$, find (a) hoop stress, (b) longitudinal stress, (c) hoop strain, (d) longitudinal strain, and (e) change in volume.

Solution

Referring to Fig. 2.6:

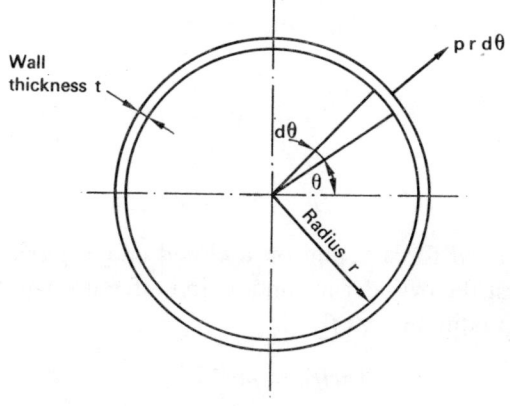

FIG. 2.6.

(a) Considering the force tending to burst the tube across a diameter and the stress set up in the material to resist it:

Radial force on an element of tube arc length
  $r \, d\theta$ at θ from a diameter $= prL \, d\theta$, where L is tube length.

Component of force acting perpendicular to
  diameter $= prL \, d\theta \sin \theta$.

Resultant force on whole area of tube above

the diameter $= \int_0^\pi prL \sin\theta \, d\theta$

$= 2prL.$

This is resisted by the stress f_H on an area of metal $2tL$.

$$\therefore 2f_H tL = 2prL,$$

$$f_H = \frac{pr}{t}.$$

(Notice (i) that the hoop stress f_H acts perpendicular to the diameter, (ii) that any diameter could have been chosen and would have given the same result so that f_H must act circumferentially all round the tube, and (iii) in a thin-walled cylinder, i.e. one where $t \ll r$, stress in the radial direction is ignored and the mean radius and the internal radius are taken as being equal.)

In this problem, $\quad p = 2.75 \times 10^6 \text{ N/m}^2,$

$r = 0.0375 \text{ m},$

$t = 0.0015 \text{ m}.$

$\therefore f_H = 68.75 \text{ MN/m}^2.$

(b) Total axial force acting on a closed end $= p\pi r^2$. The tendency to burst the cylinder around a circumference will be resisted by a longitudinal stress f_L,

where $\quad 2\pi r t f_L = p\pi r^2,$

$$f_L = \frac{pr}{2t}.$$

$$\therefore f_L = \frac{2.75 \times 10^6 \times 0.0375}{2 \times 0.0015},$$

$$f_L = 34.375 \text{ MN/m}^2.$$

(c) The hoop stress will be accompanied by a corresponding strain equal to f_H/E, and the longitudinal stress will be accompanied

STRESS AND STRAIN

by a corresponding strain f_L/E. Each of these strains will give rise to lateral strains in perpendicular directions and the total hoop strain e_H will be

$$e_H = \frac{f_H}{E} - v\frac{f_L}{E},$$

while total longitudinal strain e_L will be

$$e_L = \frac{f_L}{E} - v\frac{f_H}{E}.$$

Substituting values gives:

$$e_H = 300{\cdot}8 \times 10^{-6}.$$
(d) $$e_L = 85{\cdot}94 \times 10^{-6}.$$

(e) Change in internal volume = final volume − initial volume
$$= \pi(r+re_r)^2(L+Le_L) - \pi r^2 L$$
$$\doteqdot \pi r^2 L(1+2e_r+e_L) - \pi r^2 L$$

if all products of strains are neglected as being small quantities.

Change in internal volume $= \pi r^2 L(2e_r + e_L)$.

Now the radial strain $e_r = \dfrac{\text{change in radius}}{\text{original radius}}$

$$= \frac{\delta r}{r}.$$

But hoop strain $e_H = \dfrac{\text{change in circumference}}{\text{original circumference}}$

$$= \frac{2\pi(r+\delta r) - 2\pi r}{2\pi r}$$
$$= \frac{\delta r}{r}.$$

∴ hoop strain = radial strain.
∴ change in volume $= \pi r^2 L(2e_H + e_L)$
$$= 3{\cdot}04 \times 10^{-6} \text{ m}^3.$$

Problems

2.8. The member shown in Fig. 2.7 is an aluminium cylinder 50 mm diameter with a concentric hole 25 mm diameter bored in one end. Find the amount the rod shortens as the result of an axial compressive load of 200 kN. ($E = 69$ GN/m².)

[0·714 mm.]

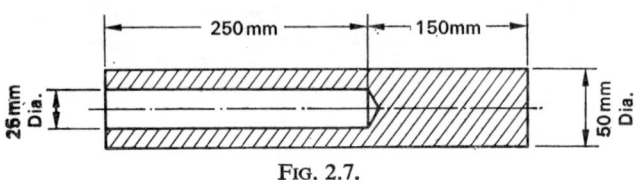

Fig. 2.7.

2.9. A bar of total length 600 mm is made up of three concentric sections: A, 150 mm long, 30 mm diameter; B, 300 mm long, 25 mm diameter; C, 150 mm long, 40 mm diameter. Find the stress in each section and the total length of the bar when a compressive axial load of 110 kN is applied to it. ($E = 200$ GN/m².)

[156 MN/m², 224 MN/m². 87·5 MN/m². 599·48 mm.]

2.10. A weight W suspended from a steel wire 4 m long, 3·25 mm diameter, causes it to lengthen by 1·69 mm. What is the value of W if $E = 200$ GN/m²? The same weight suspended from a titanium wire 3 m long, 1·75 mm diameter causes it to lengthen by 8·5 mm. Estimate the modulus of elasticity of titanium.

[700 N. $E = 102·8$ GN/m².]

2.11. A bolt used on a machine tool has an effective diameter of 5 mm and a pitch of 0·5 mm. Its length, measured between faces of nut and bolt-head, is 125 mm before tension is applied. If the nut is tightened by one-fifth of a turn, find the resulting tension in the bolt if the member in which it fits remains rigid. If the strain at yield point is 0·1 per cent, what fraction of a turn will induce yield stress in the bolt? ($E = 200$ GN/m².)

[3142 N. $\frac{1}{4}$ turn.]

2.12. Two vertical wires, both 5 m long, are suspended at a distance 0·5 m apart. At the upper ends they are attached to a rigid horizontal beam and at the lower ends they support a light but rigid horizontal bar which carries a vertical load P somewhere between the wires.

If the left-hand wire is of copper and 2 mm diameter, while the right-hand wire is of steel and 1 mm diameter, find (a) the distance of P from the left-hand wire, (b) the load, stress and extension in each wire if $P = 175$ N. ($E_{\text{steel}} = 200$ GN/m². $E_{\text{Cu}} = 110$ GN/m².)

[(a) 156 mm. (b) Copper wire: 120·4 N, 38·3 MN/m². Steel: 54·6 N, 69·5 MN/m². Extension is same in each, 0·348 mm.]

2.13. Two fine wires each having a cross-sectional area of $0·735 \times 10^{-6}$ m² and a length of 0·25 m support a load as shown in Fig. 2.8. If the wires only undergo elastic straining, find the vertical deflection of point A due to the load. ($E = 200$ GN/m².)

[0·43 mm.]

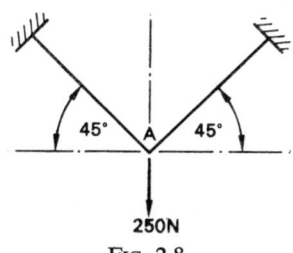

FIG. 2.8.

2.14. Three wires each having a cross-sectional area of $1·25 \times 10^{-6}$ m² and a length of 0·25 m are joined at point A (Fig. 2.9). The other ends of the wires are fastened to rigid mountings and the wires are initially taut but unstrained. If a load of 250 N is applied vertically down at point A, find the stress in each wire and the vertical deflection of A. ($E = 200$ GN/m².)

[200 MN/m². 200 MN/m². 0, 0·51 mm.]

2.15. The material used for a propeller shaft has a modulus of rigidity of 85 GN/m². A small square is scribed on the shaft surface

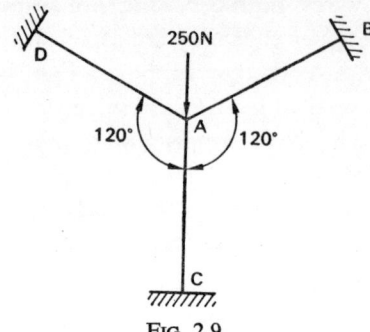

Fig. 2.9.

with two of its sides parallel with the shaft axis, find the change in the angle at the corners of the square when the shaft is subjected to a torque which produces a shear stress of 170 MN/m² on the surface of the shaft.

[0·115°.]

2.16. Figure 2.10 shows a tension rod connection assembly. The fork is made of aluminium alloy with a shank of square cross-section 25 mm side. The tie rod is steel of circular cross-section, diameter d. Calculate the safe load P which the connection can carry if the stress in the aluminium shank must not exceed 70 MN/m² and the stress in the steel rod must not exceed 100 MN/m². What is the optimum value of d?

[$P = 10·6$ kN. $d = 11·63$ mm.]

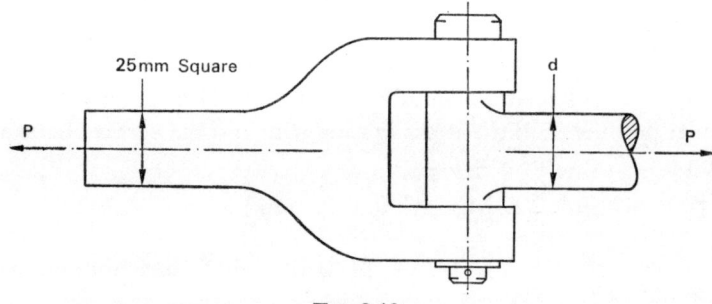

Fig. 2.10.

2.17. A bar of steel of rectangular cross-section 75 mm by 25 mm is subjected to a tensile load of 180 kN. Find the change in dimensions of the cross-section if $v = \frac{1}{3}$, $E = 200$ GN/m².

[0·012 mm. 0·004 mm. Both decrease.]

2.18. Find the change in volume of a prismatic bar 500 mm long of circular cross-section 40 mm diameter when an axial tensile load of 100 kN is applied. ($E = 200$ GN/m². $v = \frac{1}{3}$.)

[Volume increases by 83·33 mm³.]

2.19. A tensile test specimen made of aluminium alloy has a diameter of 0·0113 m and gauge length of five times the diameter. (N.B.—This ratio of gauge length/diameter is recommended in BS 18: 1962.) In a tensile test to destruction, the following results were obtained:

Load (kN) 2 4 6 8 10 12 14 16 18 20 22 24
Extension
(μm)   1·6 3·3 5·0 6·6 8·0 10·0 11·5 13·5 18·0 25·0 36·0 57·0

Plot these values and find from the graph (a) modulus of elasticity, and (b) 0·05 per cent and 0·1 per cent proof stresses.

[(a) $E = 69.5$ GN/m². (b) 231 MN/m²; 247 MN/m².]

2.20. A tensile test on a specimen of 0·13 per cent carbon steel gave the following results: gauge length 56·5 mm, diameter 11·3 mm; gauge length after fracture 73·5 mm.

Plot two graphs of load/extension (1) up to a load 25·0 kN, (2) for the whole test, and find (a) modulus of elasticity, (b) stress at yield point, (c) ultimate tensile stress, (d) percentage elongation at fracture, and (e) percentage strain at yield point.

Load (kN)	2·5	5·0	7·5	10	12·5	15	17·5	20	22·5
Extension (μm)	6·5	14	20	28	34	42	48	56	62
Load (kN)	25	26	30	35	37·5	39	38	36	28
Extension (μm)	70	100	250	560	740	1000	1200	1400	1700

[(a) 204 GN/m². (b) 250 MN/m². (c) 395 MN/m². (d) 30·1 per cent. (e) 0·1225 per cent.]

2.21. A tensile test piece of 0·29 per cent carbon steel was made to the following dimensions: gauge length 28·2 mm, diameter 5·6 mm. The results of the test are given below:

Load (kN)	1	2	3	4	5	6	7	7·7	8
Extension (μm)	5	11	16	21·5	27	32·5	38	43	48

Show that this material obeys Hooke's Law and find the modulus of elasticity of the material. Estimate the stress at yield point and at the limit of proportionality.

[$E = 210$ GN/m². $f_{YP} = 308$ MN/m². $f_{PL} = 256$ MN/m².]

2.22. A steel bar of 20 mm diameter is 300 mm long. A brass tube of 25 mm outside diameter fits loosely over the steel bar and is of the same length. The combined bar stands vertically on a solid horizontal base and supports an axial compressive load of 50 kN. Find the stress in the bar and the tube and the compression caused by the load. ($E_{bar} = 200$ GN/m². $E_{tube} = 75$ GN/m².)

[131·4 MN/m². 49·3 MN/m². 0·197 mm.]

2.23. A steel bar 22 mm diameter and 500 mm long stands vertically on a rigid horizontal base. A brass tube 25 mm internal diameter, 40 mm external diameter, and 500·25 mm long stands concentrically around it. A rigid cover plate is placed over the combined column and a compressive axial load of 100 kN is applied. Find the compressive stresses in (a) the steel bar, and (b) the brass tube. ($E_s = 200$ GN/m². $E_B = 100$ GN/m².)

[(a) 80·92 MN/m². (b) 90·4 MN/m².]

2.24. A cylindrical steel rod of cross-sectional area 650 mm² and length 99·95 mm is surrounded concentrically by a loose-fitting brass tube 100 mm long. Both ends of the tubes are closed by perfectly stiff plates to which an axial compressive force of 150 kN is applied. If the stress in the steel is not to exceed 125 MN/m² and the stress in

STRESS AND STRAIN

the brass is not to exceed 90 MN/m², find the required cross-sectional area of the brass tube. ($E_s = 200$ GN/m², $E_B = 100$ GN/m².)

[London Univ.]

[The brass condition is the critical one and gives area = 1083 mm².]

2.25. A cast-iron tube 900 mm long, 150 mm internal diameter, and 12·5 mm thick has each end closed by a 50 mm thick rigid cover plate. The plates are secured by means of a 25 mm diameter high tensile steel bolt located centrally in the bore of the tube. The bolt is initially tensioned until a strain gauge placed on the outer surface of the tube records an axial strain of $0·5 \times 10^{-3}$. The bolt is subsequently subjected to an external tensile force. Determine a relationship between the stress in the bolt and the external force. Find the external tensile force on the bolt that results in zero interface pressure between tube and cover plates.

[London Univ., B.Sc. 1, Prop. Mat. and Stress Analysis, 1968]

[If external load = P newtons and total stress in bolt = f_B, $f_B = 2040P + (570 \times 10^6 - 1·24 \times 10^4 P)$ for P less than 46 kN. Force for zero interface pressure = 46 kN and above this value $f_B = 2040P$.]

2.26. A steel bar 4 m long, 50 mm diameter, has its temperature raised by 80°C. It is then found to fit perfectly between two blocks to which it is securely fastened. If the bar is then cooled to its original temperature and the blocks remain rigid, what will be the force exerted on the blocks and the stress in the bar? ($\alpha = 0·000011/°C$. $E = 200$ GN/m².)

[345·4 kN. 176 MN/m².]

2.27. For the same situation as problem 2.26, what will be the force exerted on the blocks and the stress in the bar if the blocks do not remain rigid but move towards one another such that their distance apart decreases by 1 mm when the bar cools?

[247·4 kN. 126 MN/m².]

2.28. A steel tie-rod of 25 mm diameter is placed concentrically in a brass tube of 60 mm bore and 68 mm outside diameter. Nuts and washers are fitted on the tie-rod so that the washers completely cover the ends of the tube. The nuts are tightened to give a compressive stress of 30 MN/m² in the tube. Find the strains in the rod and tube and the stress in the rod. ($E_{steel} = 200$ GN/m². $E_{brass} = 80$ GN/m².)

[0.305×10^{-3} in rod. 0.375×10^{-3} in tube. 61.5 MN/m².]

2.29. For the situation of problem 2.28 suppose a tensile load of 45 kN is applied to the tie rod, find the total stresses in both tie-rod and tube resulting from the combined prestress and external load (a) if there is no temperature change, (b) if the temperature increases by 60 °C. ($\alpha_{steel} = 11 \times 10^{-6}$/°C. $\alpha_{brass} = 18.9 \times 10^{-6}$/°C.)

[London Univ.]

[(a) 91.6 MN/m² in rod, zero in tube. (b) 122.7 MN/m² in rod, 18.98 MN/m² compressive in tube.]

2.30. What working pressure may be allowed in a cylindrical boiler 2.5 m internal diameter with plates 20 mm thick if the maximum tensile stress allowed in the plates is 70 MN/m². What will be the hoop and longitudinal stresses in the shell under these conditions?

[1.12 MN/m². 70 MN/m². 35 MN/m².]

2.31. A boiler drum consists of a cylinder with flat ends; the length is 2.5 m, diameter 1.2 m, wall thickness 25 mm. It is filled completely full with water and additional water is pumped in until the pressure rises to 7 MN/m². What extra volume of water is needed at this pressure? ($E_{steel} = 200$ GN/m². $\nu = 0.3$. $K_{water} = 2.21$ GN/m².)

[13.4×10^{-3} m³.]

2.32. An aluminium cylinder 2.5 mm thick is strengthened by having a steel cylinder also 2.5 mm thick shrunk on to it. Initially there is an interference fit of 0·1 mm measured on the diameter. The mean diameter of the combination is approximately 100 mm. Find the hoop stresses in each "thin shell" caused by the interference fit. ($E_{steel} = 210$ GN/m². $E_{al} = 70$ GN/m².)

[52·5 MN/m² tensile in steel; compressive in aluminium.]

2.33. In a tensile test on a steel tube of external diameter 0·020 m and 0·014 m bore, an axial load of 2 kN produced a stretch of 3 μm on a 0·05 m gauge length, and a lateral contraction of 0·3 μm on the outside diameter. Calculate E, v, and K for the material.
[$E = 208 \cdot 3$ GN/m², $v = 0 \cdot 25$. $K = 138 \cdot 8$ GN/m².]

2.34. A round bar is suspended vertically from a rigid support. At the bottom of the bar is a rigid collar which arrests the fall of an annular weight of 1000 N after it has dropped a distance of 0·2 m. If the bar is 6 m long what must its diameter be in order that the stress due to the impact does not exceed 80 MN/m². ($E = 200$ GN/m².)

[51 mm.]

2.35. For the system described in problem 2.34 suppose that the falling body has a weight W, that it falls a distance h, and that the bar is of length L, cross-sectional area A, and elastic modulus E. Derive a relation for the maximum stress reached in the bar due to impact, and by using the special case of h becoming vanishingly small show that the stress due to a suddenly applied—but not dropped—load is twice the stress produced by the same load under static conditions.

2.36. A mass of 1000 kg is being steadily lowered by a crane at a speed of 0·4 m/s. The crane cable is a wire rope of diameter 25 mm ($E = 200$ GN/m²) and at the moment when the length of cable between the load and the pulley at the tip of the crane jib is 10 m the pulley seizes and jams the cable. Find (a) the stress in the cable while the load is being lowered steadily, and (b) the maximum stress reached in the cable at the instant the cable jams. ($g = 9.81$ m/s².)

[20 MN/m². 100·6 MN/m², i.e. 80·6 due to the sudden stop and 20 due to the weight of the load.]

2.37. A weight of 100 N is supported by a vertical mild steel wire of 2 mm diameter 10 m long. A workman inadvertently allows the weight to fall freely for a distance of 6 mm before the wire becomes taut. Assuming the wire deforms elastically and that no energy is lost, calculate the maximum stress reached in the wire. ($E = 200$ GN/m²,)

[124·8 MN/m².]

2.38. A vertical steel rod of 25 mm diameter checks the fall of a weight of 2000 N which strikes it axially after falling 3·8 mm. Assuming the rod does not buckle, what is the shortest length of rod that will bear the impact without the stress in it exceeding 125 MN/m².

[425 mm.]

2.39. A bar of mild steel 2 m long is subjected to a tensile load of 8 kN. The cross-section of the bar is 500 mm and the modulus of elasticity is 200 GN/m². Find (i) the stress in the bar, (ii) the extension under this load, (iii) the strain energy stored in the bar, and (iv) the strain energy that would be stored in the bar if half its length were reduced in diameter to give half the original cross-sectional area, while the load remains at 8 kN.

(i) 16 MN/m². (ii) 160 μm. (iii) 0·64 J. (iv) 0·8 J.]

2.40. Calculate the modulus of resilience for the materials listed below and arrange them in descending order of magnitude. If the modulus were "per unit mass" instead of "per unit volume", what would the order then be?

(a) 0·2 per cent carbon steel, cold rolled. ($E = 200$ GN/m², $f_{EL} = 414$ MN/m², $\varrho = 7820$ kg/m³.)

(b) Nickel steel, oil quenched. ($E = 200$ GN/m², $f_{EL} = 1103$ MN/m², $\varrho = 7820$ kg/m³.)

(c) Hard drawn copper. ($E = 117$ GN/m², $f_{EL} = 262$ MN/m², $\varrho = 8900$ kg/m³.)

(d) Hard drawn aluminium. ($E = 69$ GN/m², $f_{EL} = 138$ MN/m², $\varrho = 2680$ kg/m³.)

[(a) 427 kJ/m³. (b) 3050 kJ/m³. (c) 294 kJ/m³. (d) 138 kJ/m³. Per unit mass (d) comes before (c) since (d) has 51·5 J/kg while (c) has 33 J/kg.]

CHAPTER 3

Two-dimensional Stress Systems

Definitions and Theory

(a) If shear stresses of a particular size and direction exist on a certain plane in a material, then on a perpendicular plane there will be complementary shear stresses equal in magnitude and opposite in direction.

(b) If an axial tensile force P is applied to a bar of regular cross-sectional area A, the stresses on a plane at θ to the bar axis will be a direct stress f and a shear stress q (Fig. 3.1), where $f = P/A \sin^2 \theta$, $q = \frac{1}{2} P/A \sin 2\theta$.

(c) *Stress sign conventions.* The following conventions are those used in most standard textbooks. Direct stresses: tensile positive: Compressive negative. Shear stresses: general convention considers shear

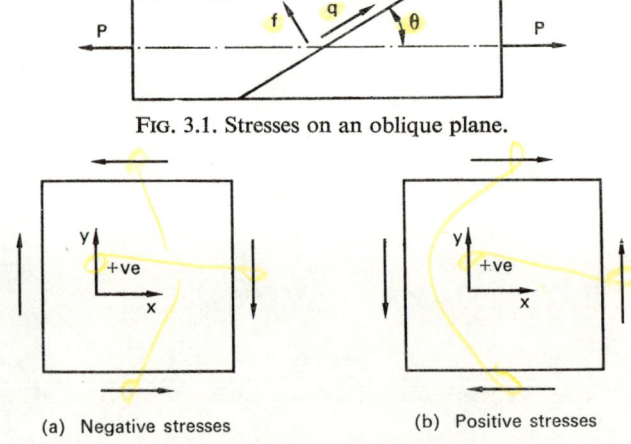

FIG. 3.1. Stresses on an oblique plane.

(a) Negative stresses   (b) Positive stresses

FIG. 3.2. General shear stress sign convention.

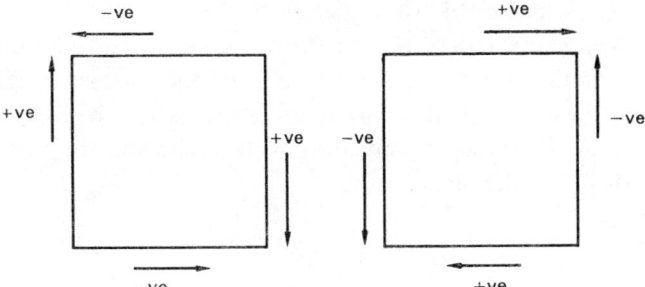

FIG. 3.3. Shear stress convention for Mohr's circle.

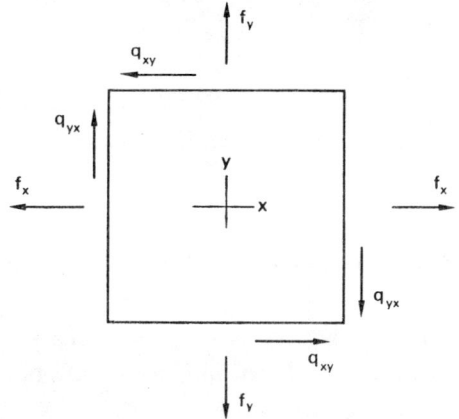

FIG. 3.4. Stress symbols.

stresses positive if the direction of the stress and the direction of the normal to the plane on which the stress acts both have the same sign relative to the coordinate axes of the system. If the directions have different signs, the stress is considered negative. The shear stress convention is illustrated in Fig. 3.2.

For Mohr's circle the special convention is used that a shear stress is plotted positive on the circle diagram if it appears to have a clockwise-turning moment relative to the face on which it acts; it is plotted negative if it appears to have a counter-clockwise moment. Figure 3.3 shows the convention for Mohr's circle.

(d) The most general two-dimensional stress system is illustrated in Fig. 3.4. Suffix letters on the direct stresses f indicate both the direction of the stress and the direction of the normal to its plane relative to the conventional coordinate directions. On the shear stresses the first suffix indicates the direction of the stress and the second that of the normal to the plane.

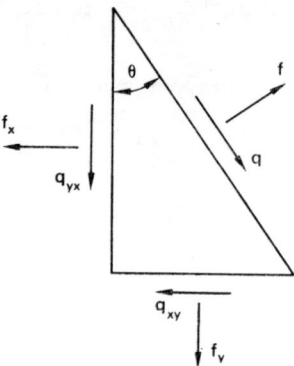

Fig. 3.5. Stresses in a two-dimensional system.

(e) The direct stress f and the shear stress q on a plane inclined at θ to the y axis in the system shown in Fig. 3.5 are obtained from the equations:

$$f = \tfrac{1}{2}(f_x+f_y)+\tfrac{1}{2}(f_x-f_y)\cos 2\theta + q_{xy} \sin 2\theta,$$
$$q = \tfrac{1}{2}(f_x-f_y)\sin 2\theta - q_{xy} \cos 2\theta.$$

For derivation, see example 3.3.

(f) Mohr's stress circle is a graphical construction which allows the direct and shear stresses for any value of θ to be determined graphically. Shear stresses are plotted vertically and direct stresses horizontally, and the stresses on any face of an elemental system such as that shown in Fig. 3.5 can be represented by the coordinates of a point relative to two axes. Figure 3.6 shows Mohr's circle for the system of Fig. 3.5 on the assumption that $f_x > f_y$. The circle is drawn by locating the points which represent the stresses on the x and y

faces, joining these points and using the line thus obtained as the basic diameter of a circle. The stresses on a plane at an angle θ measured counter-clockwise from the x face of the element are the coordinates of a point on the circumference of the circle which lies at 2θ counter-clockwise from the x point at the end of the basic diameter.

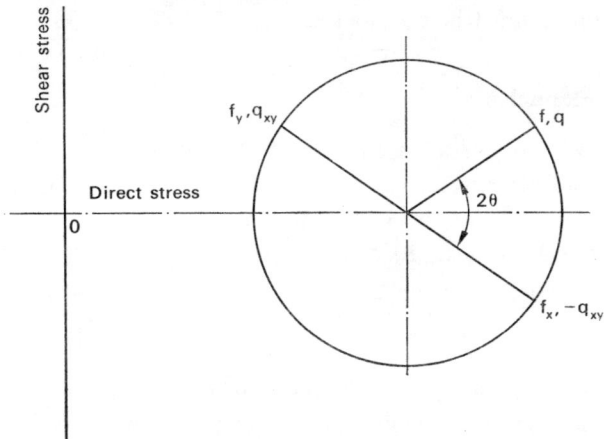

FIG. 3.6. Mohr's circle for the system shown in Fig. 3.5.

(g) No matter what the stresses acting on an element such as that shown in Fig. 3.4, it will always be possible to find two mutually perpendicular planes through the point on which the shear stresses are zero and on which the direct stresses are maximum and minimum. These planes are called the "principal planes". In a three-dimensional stress system there would be three of these planes each carrying a normal stress, f_1, f_2, or f_3. In many practical problems, though the material carrying the stress has three dimensions, it is only subjected to two-dimensional stresses, that is $f_3 = 0$. It is clear from Fig. 3.10, example 3.4, that the maximum shear stress is equal to the radius of Mohr's circle, i.e. $q_M = \frac{1}{2}(f_1-f_2)$. Now the plane for which Mohr's circle is drawn can be chosen such that it is the plane of f_1 and f_3 and $q_M = \frac{1}{2}(f_1-f_3)$, where $f_3 = 0$ for a two-dimensional system. If f_1 and f_2 are of different signs $\frac{1}{2}$, (f_1-f_2) will be greater than $\frac{1}{2}(f_1-0)$

and should be quoted as the maximum shear stress, but if f_1 and f_2 are of the same sign, the overall maximum shear stress will be $\frac{1}{2}(f_1-0)$.

In some problems the stress system is specified as being "on a plane element", and the intention then is that all effects outside the plane should be ignored. If, however, the problem refers to a three-dimensional element, the possibility of a maximum shear stress being in another plane must be considered.

Worked Examples

3.1. A steel bar carries a torque which causes the maximum shear stress, at the bar surface, to be 80 MN/m². What will be the normal and shear stresses at the bar surface on planes inclined at (a) 30°, (b) 45°, (c) 60°, and (d) 135° to the cross-sectional plane?

Solution

The method consists of considering a small element of material surrounding a point on the bar surface. The element is of finite size so that the stresses can be multiplied by the areas on which they act to produce forces that can be included in the equilibrium equations for the element. In the course of the analysis the dimensions of the element disappear and the result is true for the stress at a point. Refer to Fig. 3.7.

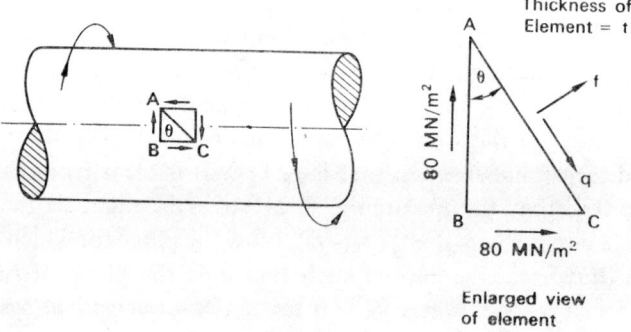

FIG. 3.7.

Resolving forces perpendicular to AC, outwards positive.

$$f(AC)\,t + 80(AB)\,t\sin\theta + 80(BC)\,t\cos\theta = 0,$$
$$f = -80(AB/AC)\sin\theta - 80(BC/AC)\cos\theta,$$
$$f = -80[(\cos\theta\sin\theta) + (\sin\theta\cos\theta)],$$
$$f = -80\sin 2\theta.$$

Resolving forces parallel to AC, downwards positive.

$$q(AC)\,t - 80(AB)\,t\cos\theta + 80(BC)\,t\sin\theta = 0,$$
$$q = 80\cos^2\theta - 80\sin^2\theta,$$

(a) $\theta = 30°$.

$$f = -80\sin 60°$$
$$= -40\sqrt{3}\ \text{MN/m}^2$$
$$= 69\cdot3\ \text{MN/m}^2\ \text{compression}.$$
$$q = 80(\tfrac{3}{4} - \tfrac{1}{4})$$
$$= 40\ \text{MN/m}^2.$$

(b) $\theta = 45°$.

$$f = 80\ \text{MN/m}^2\ \text{compression}.$$
$$q = 0.$$

(c) $\theta = 60°$.

$$f = 69\cdot3\ \text{MN/m}^2\ \text{compression}.$$
$$q = -40\ \text{MN/m}^2.$$

(d) $\theta = 135°$.

$$f = 80\ \text{MN/m}^2\ \text{tension}.$$
$$q = 0.$$

3.2. A tie-rod is to be made by welding, end to end, two short bars of the same rectangular cross-section. At what angle to the bar axis should the joint be made if (a) the joint can carry the same size stress in shear as it can in tension and it is desired to make the joint as strong as possible, and (b) the joint is not to carry any shear stress?

The correctness of the answers is to be shown from first principles.

Solution

Refer to Fig. 3.8.

Considering the free-body diagram of the part of the bar above the joint and assuming that the cross-sectional area is A.

$$\text{Normal force} = P \cos \theta.$$
$$\text{Shear force} = P \sin \theta.$$
$$\text{Surface area of joint} = A/\cos \theta,$$
$$f = (P/A) \cos^2 \theta,$$
$$q = (P/A) \sin \theta \cos \theta.$$

For part (a), $\qquad f = q.$

$$\therefore \cos^2 \theta = \sin \theta \cos \theta,$$
$$\tan \theta = 1,$$
$$\theta = 45°.$$

For part (b), $\qquad q = 0.$

$$\therefore (P/A) \sin \theta \cos \theta = 0.$$

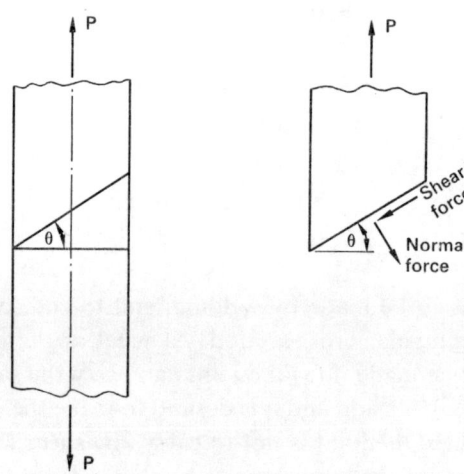

Fig. 3.8.

This can happen in three ways:
(i) $P = 0$, which means there is no load.
(ii) $\cos \theta = 0$, which means $\theta = 90°$ and the joint is parallel to the axis which is useless for lengthening a bar.
(iii) $\sin \theta = 0°$ and $\theta = 0°$. This is the required angle.

3.3. A block of material of unit thickness has acting upon it the two-dimensional system illustrated in Fig. 3.9. Find an expression for the direct stress f and the shear stress q acting on planes which make an angle θ with the x face. No graphical or algebraic relations are to be assumed.

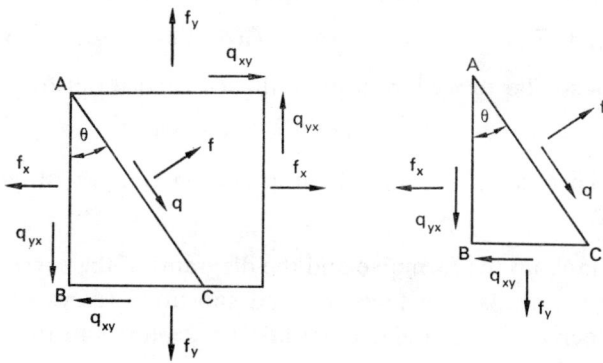

Fig. 3.9.

Solution

Draw the section ABC as a free-body diagram but show stresses instead of forces. If the stresses are converted to forces by multiplying them by the areas on which they act, then the equilibrium equations can be used for the free-body.

Sum of forces perpendicular to $AC = 0$ positive outwards.

Area of each face such as AC will be numerically equal to the length of the face since the block is of unit thickness.

$\therefore\ f(AC) - f_x(AB)\cos\theta - q_{xy}(BC)\cos\theta - q_{yx}(AB)\sin\theta$
$\quad - f_y(BC)\sin\theta = 0,$ where (AC) means "the length of AC".

Now q_{xy} is numerically equal to q_{yx}.

$$\therefore f = f_x(AB)/(AC)\cos\theta + q_{xy}(BC)/(AC)\cos\theta$$
$$+ q_{xy}(AB)/(AC)\sin\theta + f_y(BC)/(AC)\sin\theta = 0.$$

Substituting trigonometric ratios for $(AB)/(AC)$, etc.,

$$f = f_x\cos^2\theta + f_y\sin^2\theta + 2q_{xy}\sin\theta\cos\theta,$$

but $\cos^2\theta = \tfrac{1}{2}(1+\cos 2\theta)$ and $\sin^2\theta = \tfrac{1}{2}(1-\cos 2\theta).$

$$\therefore f = \tfrac{1}{2}f_x(1+\cos 2\theta) + \tfrac{1}{2}f_y(1-\cos 2\theta) + q_{xy}\sin 2\theta.$$
$$f = \tfrac{1}{2}(f_x+f_y) + \tfrac{1}{2}(f_x-f_y)\cos 2\theta + q_{xy}\sin 2\theta.$$

Resolving forces parallel to AC, downwards positive,

$$q(AC) + q_{yx}(AB)\cos\theta + f_y(BC)\cos\theta - f_x(AB)\sin\theta - q_{xy}(BC)\sin\theta = 0.$$

Using a similar procedure to that already carried out for f

$$q = \tfrac{1}{2}(f_x-f_y)\sin 2\theta - q_{xy}\cos 2\theta.$$

Note: There are four methods of attack on complex stress system problems:

(1) To look up the formulae and the diagrams of the corresponding stress systems in a textbook and substitute values of stresses.
(2) To derive the formulae from first principles as in this problem and substitute values.
(3) To solve the problem from first principles using values of stresses instead of symbols.
(4) To use Mohr's circle method.

It is not recommended that students memorize the formulae just derived since the signs in them are critical and refer to the directions of stresses assumed in the stress system and to the location of the angle θ. Some textbooks take θ as measured from the y face, and this produces valid equations with signs different from those given here.

3.4. A torque applied to a shaft produces a maximum shear stress in it of 75 MN/m². At the same time an axial tensile load produces a direct stress of 110 MN/m². Draw Mohr's circle for the system if the

TWO-DIMENSIONAL STRESS SYSTEMS

direction of the torque is such that the stress on the x face plots positive on the circle diagram and if the axial stress also acts on the x face.

From the circle find: (a) values of maximum and minimum principal stresses, (b) value of maximum shear stress, and (c) angles between shaft axis and planes carrying principal stresses and maximum shear stresses.

Solution

Mohr's stress circle is shown in Fig. 3.10.
From the circle:

Maximum principal stress $= f_1 = 148 \cdot 1$ MN/m² tensile.
Minimum principal stress $= f_2 = 38 \cdot 1$ MN/m² compressive.
Maximum shear stress $\quad = q_m = 93 \cdot 01$ MN/m².

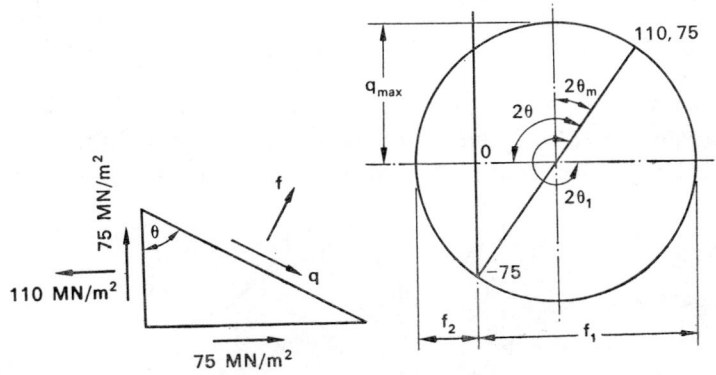

Fig. 3.10.

Angle to plane of maximum principal stress measured from x face
$$= 2\theta_1 = 306° \ 16'.$$
$$\therefore \ \theta_1 = \frac{306° \ 16'}{2} = 153° \ 08' \text{ counter-clockwise.}$$

Angle to plane of maximum shear stress measured from x face
$$= 2\theta_m = 36° \ 16'.$$
$$\therefore \ \theta_m = 18° \ 08'.$$

But angles are required relative to the bar axis which is at 90° to the x face.

Therefore relative to the axis the plane of maximum principal stress is 63° 08′ counter-clockwise and the plane of maximum shear is 108° 08′ counter-clockwise.

Note that the angles could have been obtained directly from Mohr's circle by measuring from the y end of the basic diameter.

3.5. Direct stresses of 120 MN/m² tension and 90 MN/m² compression are applied to an elastic material at a certain point on planes at right angles to each other. The greater principal stress in the material is limited to 150 MN/m². To what shearing stress may the material be subjected on the given planes and what will then be the maximum shearing stress at the point? [London Univ. B.Sc. 2]

Solution

Considering Mohr's circle, the coordinates of direct stress on the two given planes may be located, but the shear stresses on these planes are unknown. The centre of Mohr's circle lies on the direct stress

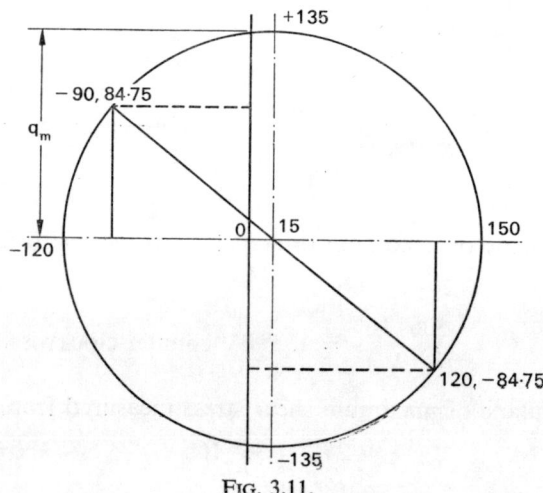

Fig. 3.11.

axis midway between the two points just located; the centre is therefore at point 15, 0. The greater principal stress will be the point where the circumference crosses the direct stress axis and the radius is therefore $150 - 15 = 135$.

The circle can now be drawn to scale and the shear stresses on the planes can be read off (see Fig. 3.11).

Shear stresses on the planes = ± 84.75 MN/m².
Maximum shear stress = 135 MN/m².

Problems

3.6. A bar of square cross-section 50 mm by 50 mm is to be lengthened by splicing-on a similar bar with adhesive. The angle of the joint is to be such that the maximum stresses in shear and tension, 4·2 MN/m² and 7 MN/m² respectively, are reached simultaneously. What should be the angle between the joint and the plane perpendicular to the bar axis? What is the maximum axial tensile load the bar can carry?

[30° 55′. 23·6 kN.]

3.7. A high tensile steel bar carries a tensile load of 10 kN; its cross-sectional area is 130 mm². Find the maximum shear stress q_{max} and the normal stress which acts on the plane of maximum shear.

[38·5 MN/m². 38·5 MN/m².]

3.8. A straight bar has a constant cross-section of 100 mm² and carries an axial load of 20 kN. Find the normal and shear stresses on planes inclined at the following angles to the bar axis: (a) 30°, (b) 45°, (c) 60°.

[(a) 50 MN/m²; 86·6 MN/m². (b) 100 MN/m²; 100 MN/m². (c) 150 MN/m²; 86·6 MN/m².]

3.9. Cast iron is weaker in tension than in shear, and weaker in shear than in compression. From this statement deduce the angle to the axis of the line of fracture of cylindrical cast iron specimens when they fracture under: (a) axial tension, (b) axial compression, and (c) torsion.

[(a) 90° to the axis. (b) 45° to the axis. (c) 45° to the axis.]

3.10. At a point within a piece of material the stress on one of the principal planes is 75 MN/m² tensile; on the other principal plane the stress is 60 MN/m² compressive. Find the normal and shear stresses on a plane which makes an angle of 30° with the principal plane first mentioned. Indicate clearly the directions of the normal and shear stresses. [London Univ., B.Sc. 1]

[The answer is given in Fig. 3.12.]

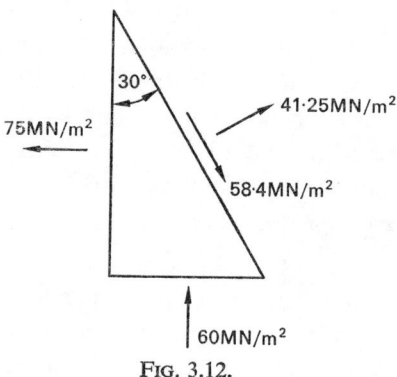

Fig. 3.12.

3.11. The principal stresses at a point in a material are 45 MN/m² and 75 MN/m², both tensile. Working from first principles, determine for a plane inclined at 40° to the plane on which the 75 MN/m² stress acts, (a) the magnitude and direction of the resultant stress, and (b) the normal and tangential components of the stress.

[London Univ., B.Sc. 2]

[(a) 64·2 MN/m² at 76° 45′ to the plane. (b) $f = 62·58$ MN/m²; $q = 14·75$ MN/m².]

3.12. If a body is subjected to tensile stresses of intensity p_1 and p_2 at right angles to one another, show how to find the normal and

TWO-DIMENSIONAL STRESS SYSTEMS 73

tangential stresses on a plane face inclined at an angle θ to the line of stress p_1.

A steel plate is subjected to tensile stresses of 120 MN/m² and 80 MN/m² at right angles to each other. Determine the normal and tangential stresses on a plane inclined at 60° to the 120 MN/m² stress.

[London Univ., B.Sc.]

$[f = 110$ MN/m². $q = 17.32$ MN/m².]

3.13. Illustrate, on appropriate sketches of the Mohr circle for stress, the following: (a) the necessity for complementary shear stress, (b) a "hydrostatic" stress system, (c) the state of stress at a point in a bar which is subjected to a tensile force only, and (d) the state of stress at a point in a shaft which is subjected to twisting only.

[London Univ., B.Sc. 1]

3.14. The loads applied to a piece of material cause a shear stress of 60 MN/m² together with a normal tensile stress on a certain plane. On a plane perpendicular to this one there is no direct stress. Treating this as a two-dimensional stress problem, find the value of the normal tensile stress if it makes an angle of 30° with the major principal stress. What are the values of the principal stresses?

[69·3 MN/m². 104 MN/m². −34·7 MN/m².]

3.15. At a point in the web of a loaded girder the longitudinal tensile stress is 80 MN/m² and the shear stress is 40 MN/m². Find the magnitude and direction of the principal stresses at the point: give the direction relative to the direction of the longitudinal tensile stress. What is the magnitude and the direction of the greatest shear stress?

[London Univ., B.Sc.]

$[+96.56$ MN/m² at $67\frac{1}{2}°$. -16.56 MN/m² at $157\frac{1}{2}°$, 56.56 MN/m² at $112\frac{1}{2}°$.]

3.16. A cylindrical boiler 2 m diameter has a wall thickness of 15 mm and is subjected to an internal pressure of 180 kN/m². Assuming that the theory for a thin cylinder with closed ends applies, find the values of the maximum and minimum principal stresses and the maximum shear stress in the curved walls of the boiler.

[12 MN/m². 6 MN/m². 6 MN/m².]

3.17. A compressive stress of 48 MN/m² acts on a certain plane in a material and a tensile stress of 32 MN/m² acts on a perpendicular plane. The shear stresses q_{xy} and q_{yx} are also present and have the

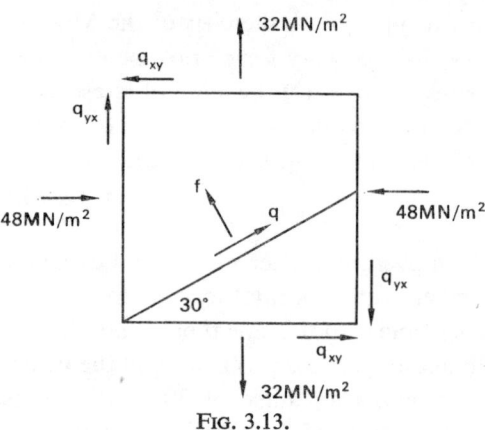

Fig. 3.13.

directions shown in Fig. 3.13. Find the value the shear stresses must have if the plane at 30° to the direction of the 48 MN/m² stress is to be a principal plane. Also find the principal stresses and the value and direction of the maximum shear stress.

[69·3 MN/m². 72 MN/m². − 88 MN/m², 80 MN/m² at 45° to the first principal plane counter-clockwise.]

3.18. At a certain point within a material the maximum and minimum principal stresses are +90 MN/m² and +30 MN/m² respectively. Find the normal and shearing stresses on a plane passing through the point and making an angle of $\tan^{-1} 0·25$ with the plane of the maximum principal stress. [London Univ., B.Sc.]

[86·5 MN/m². 14·1 MN/m².]

3.19. At a certain point in a piece of material there exists the two-dimensional stress system shown in Fig. 3.14. The resultant stress on plane AB is 102 MN/m² at 60° to the plane, and on plane BC there is a normal stress of 40·5 MN/m² as well as an unstated shear stress. Plane AC is a principal plane. Find the values of the two principal stresses and the corresponding values of θ, and also the magnitude of the maximum shear stress and the location of the plane on which it acts. [London Univ., B.Sc.]

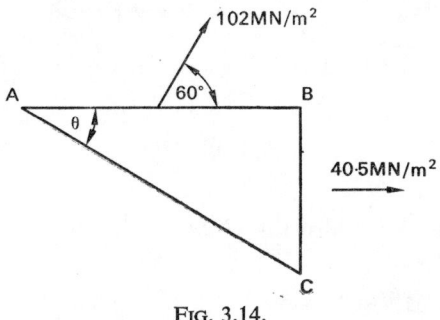

FIG. 3.14.

[$f_1 = +120\cdot75$ MN/m². $\theta_1 = 32\frac{1}{2}°$. $f_2 = +8\cdot15$ MN/m². $\theta_2 = 122\frac{1}{2}°$. $q_M = 60\cdot37$ MN/m² on a plane at 45° to the plane on which f_1 acts, and perpendicular xz plane. (N.B.—Since f_1 and f_2 are both positive, the absolute maximum shear stress is not $\frac{1}{2}(f_1-f_2)$ but $\frac{1}{2}(f_1-0)$).]

3.20. The stress system shown in Fig. 3.15a acts at a point in a solid body. Find (a) the magnitudes and directions of the principal stresses, and (b) the magnitude and direction of the maximum shear stress acting at the point.

[$f_1 = +1\cdot3$ MN/m² at 39·35° to the x face; $f_2 = -141\cdot3$ MN/m² at 129·35° to the x face. $q_M = 71\cdot3$ MN/m² at 45° to plane of f_1.]

3.21. The stress system shown in Fig. 3.15b acts at a point in a solid body. Find (a) the magnitudes and directions of the principal stresses,

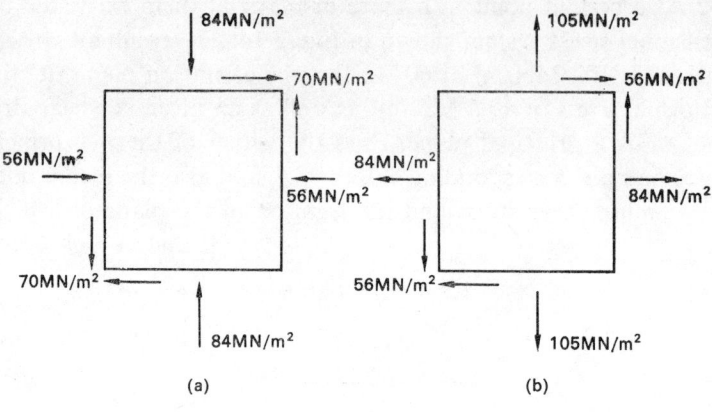

FIG. 3.15.

and (b) the magnitude and direction of the maximum shear stress acting at the point.

[$f_1 = 151{\cdot}4$ MN/m² at 39·7° to the y face. $f_2 = +37{\cdot}6$ MN/m² at 129·7° to the y face. $q_M = \frac{1}{2}(151{\cdot}4 - 0) = 75{\cdot}7$ MN/m² at 45° to f_1 relative to the plane of f_1 and f_3.]

3.22. A plane element is subjected to the stresses shown in Fig. 3.16a. Determine (a) the magnitude and direction of the principal stresses,

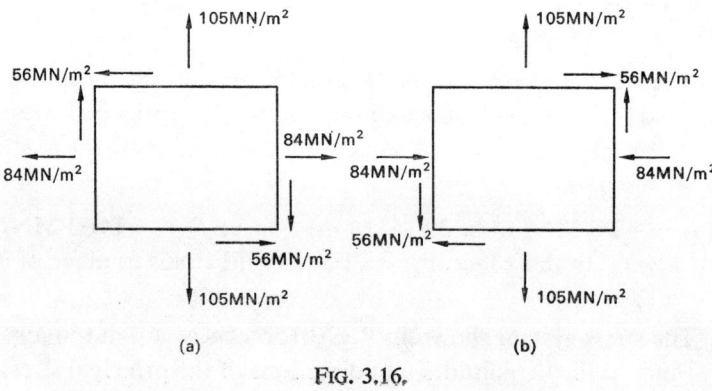

FIG. 3.16.

TWO-DIMENSIONAL STRESS SYSTEMS

(b) the magnitude and direction of the maximum shear stresses and the value of the normal stress on the plane of maximum shear, and (c) the value of the maximum shear if the figure represented a two-dimensional stress system in a solid body.

[(a) $f_1 = 151 \cdot 5$ MN/m² at 50° 28' clockwise to the x direction.
$f_2 = +37 \cdot 5$ at 39° 42' counter-clockwise to the x direction.
(b) $q_M = 57$ MN/m² at 84° 42' counter-clockwise to the x direction, $f = +94 \cdot 5$ MN/m².
(c) $q_M = \frac{1}{2}(151 \cdot 5 - 0) = 75 \cdot 75$ MN/m² $f = +75 \cdot 75$ MN/m².]

(*Note:* This question asks for the direction of the principal stresses—not the angle the principal planes make with the axes.)

3.23. Repeat problem 3.22 for the plane element illustrated in Fig. 3.16b.

[(a) $f_1 = 120 \cdot 4$ MN/m² at 105° 18' counter-clockwise to the x axis.
$f_2 = -99 \cdot 4$ MN/m² at 15° 18' clockwise to the x axis.
(b) $q_M = 109 \cdot 9$ MN/m² at 29° 42' counter-clockwise to the x axis, $f = 10 \cdot 5$ MN/m².
(c) $q_M = 75 \cdot 75$ MN/m².]

3.24. At a point in a body the stresses are a tensile stress of 84 MN/m² and a shearing stress on the same plane of 28 MN/m² negative according to the general sign convention.

(a) Sketch the stress system taking the x axis as being in the direction of the tensile stress and assuming the direct stress in the y direction is zero.
(b) Find and illustrate on separate sketches the stresses on a plane inclined at 30° to the x axis.
(c) The principal stresses.
(d) The stresses on the planes of maximum shear.

[Answer given in Fig. 3.17.]

3.25. A bar of circular cross-section diameter 50 mm carries a tensile load of 10 kN and a torque of 250 Nm in the direction shown in Fig.

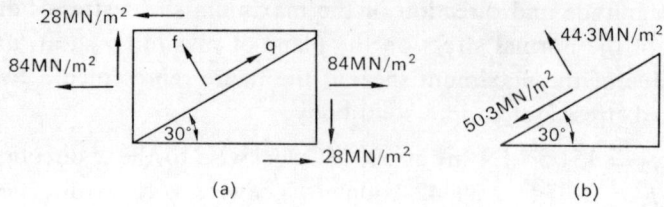

Fig. 3.17.

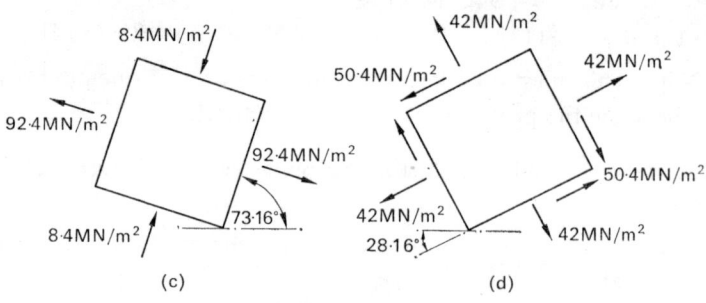

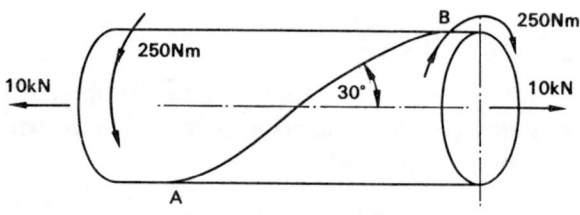

Fig. 3.18.

3.18. Find the stresses at the bar surface normal and parallel to the line AB which is marked on the surface at 30° to the bar axis.

[$f = +10 \cdot 03$ MN/m². $q = -7 \cdot 3$ MN/m².]

3.26. At a certain point in a piece of elastic material there are normal stresses of 45 MN/m² tensile and 30 MN/m² compressive on two planes at right angles to one another, together with shear stresses on the same planes. If the loading on the material is increased so that the

TWO-DIMENSIONAL STRESS SYSTEMS

stresses reach values K times those given, determine the value of K if the maximum direct stress is not to exceed 120 MN/m² and the maximum shear stress is not to exceed 75 MN/m².

[London Univ., B.Sc. 2]

[Direct stress condition $K = 2\cdot34$. Shear stress condition $K = 1\cdot72$.]

3.27. A thin walled cylindrical pipe 0·7 m diameter, wall thickness 3 mm has closed ends and carries an internal pressure of 206 kN/m² together with an axial tensile load of 230 kN. Find the maximum tensile stress in the wall of the pipe and the maximum shear stress.

[$f_{max} = 47$ MN/m². $q_{max} = 23\cdot5$ MN/m² {i.e. $\frac{1}{2}(f_1-0)$}.]

3.28. A solid shaft is carrying torque and a tensile axial load. If the maximum shear stress in the shaft due to the torque is 100 MN/m² and the shaft diameter is 100 mm, what is the maximum value that the tensile load can be without either the maximum shear stress due to the complex stress system exceeding 200 MN/m² or the maximum principal stress exceeding 400 MN/m²?

[2·72 MN.]

3.29. A bar of steel is loaded so that at a point on its surface there is a tensile stress parallel to the bar axis of 200 MN/m² and a shear stress in the same direction of 100 MN/m². Draw an element of material showing the stresses, write down the appropriate formulae for the principal stresses, and calculate from the formulae the values of these stresses. Compare these values with those obtained from solving the problem using Mohr's circle.

[$f_1 = 241\cdot4$ MN/m². $f_2 = -41\cdot4$ MN/m².]

3.30. Sketch the simplest two-dimensional stress system which would give rise to Mohr's circle of the form of (a) a point whose coordinatee are (+100, 0), (b) a point whose coordinates are (−100, 0), (c) a circle centre (0, 0), diameter 100, (d) a circle centre (50, 0), diameter 100, (e) a circle centre (− 50, 0), diameter 100.

(*Note:* There is only a single correct solution to (a) and to (b); (d) and (e) have many possible solutions but in each case one is far simpler than any other; (c) has many solutions, two of which are much simpler than the others.)

CHAPTER 4

Stresses in Beams

Definitions and Theory

(a) A beam is a long, slender member carrying loads acting perpendicular to its length or moments about axes perpendicular to its length. Beam supports are usually either "simple", i.e. they do not prevent the beam from rotating, or "built in", i.e. all rotation is prevented. A "simply supported" beam has two "simple" supports; a "built-in" or "encastré" beam is built in at both ends; a "cantilever" has a single built-in support—the other end is unrestrained. Loads on a beam are imagined as being either distributed over some length or concentrated at a point.

(b) If part of a beam is drawn as a free-body, as in Fig. 4.1, it is clear that to maintain equilibrium it may be necessary for a force and a moment to be exerted on the free-body by the material of the

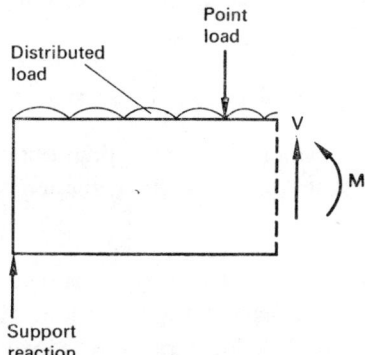

FIG. 4.1. Shear force V and bending moment M.

rest of the beam. The force gives rise to shearing stress and the moment to tensile and compressive "bending" stresses in the material. In order to calculate these stresses it is first necessary to determine the force (shear force, symbol V) and the moment (bending moment, symbol M) causing them.

(c) The shear force V at any section of a beam is defined as the sum of the transverse forces on one side only of the section. It does not matter which side, but if the summation is done on the left then upward forces are given a positive sign and downward forces negative; if the summation is done on the right, downward forces are positive and upward negative.

(d) The bending moment M at any section is defined as the sum of the moments taken about the section of all the forces on one side only. The sign convention is that any force which tends to cause sagging at the section is said to have a positive bending moment, and any force which tends to cause hogging is said to have a negative bending moment (Fig. 4.2).

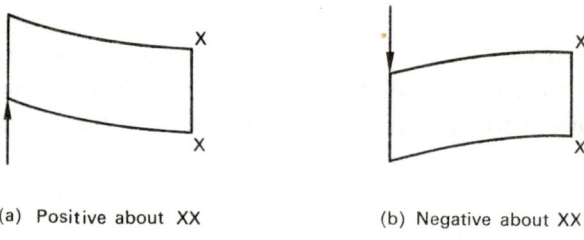

(a) Positive about XX (b) Negative about XX

FIG. 4.2. Bending-moment sign conventions.

(e) Shear force and bending moment diagrams are graphs showing the distribution of these quantities along the length of a beam.

(f) It can be shown that there is a relation between the shear force V and the bending moment M at any cross-section such that $V = dM/dx$, where x is the distance along the beam. This gives rise to the important fact that M is a local maximum or minimum where V changes sign by passing through zero.

(g) With certain preliminary assumptions, it can be shown that the bending moment M, second moment of area about the neutral axis I, bending stress f, vertical distance from the neutral axis y, beam material modulus E, and radius of curvature R are related by the equations

$$\frac{M}{I} = \frac{f}{y} = \frac{E}{R}.$$

(h) The distribution of shearing stress q across a section is obtained from the equation

$$q = \frac{V}{Ib} \int_{y_1}^{y_2} y \, dA,$$

where V is the shearing force, I is the second moment about the neutral axis, and y is the vertical distance from the neutral axis of the element of area dA, b is the breadth of the beam at a vertical height y_1, and $\int_{y_1}^{y_2} y \, dA$ is the first moment of area about the neutral axis of the area of cross-section of the beam above the level at which the stress is calculated.

(i) If end loads are applied along the axis of a beam which is, at the same time, undergoing bending, the total stresses in the beam may be obtained by the principal of superposition. That is, provided that the elastic limit is not exceeded, the combined stresses are equal to the sum of the stresses which would be developed if the bending effect and end loads were each acting separately.

(j) If an end load is applied to a member such that its line of action is parallel to but not coincident with the neutral layer passing through the centroid of the cross-section, it is called an "eccentric load" and

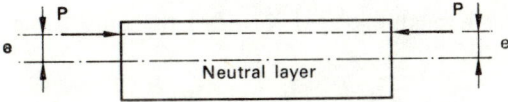

FIG. 4.3. Eccentric loading.

will give rise to combined bending and end load. The forces P (Fig. 4.3) act at a distance e from the neutral layer; they have the same effect as an axial load P coincident with the neutral layer together with moments Pe acting at each end. The stresses can then be found by superposing those for the two effects separately.

Worked Examples

4.1. Write down the equations for the bending moment and shearing force at a distance x from the left-hand end of the beams listed below and sketch the loading, bending moment, and shear force diagrams giving maximum and minimum values:

(a) Simply supported beam of span L with a point load W at $L/3$ from the left-hand support.
(b) Simply supported beam of span L with a uniformly distributed load of w N/m over the whole span.
(c) Cantilever of length L built in at left-hand end with a downward point load W at free end.
(d) Cantilever of length L built in at right-hand end with a uniformly distributed load w N/m over the whole length.

Solution

Begin by drawing the loading diagrams—the top ones in Fig. 4.4. Then apply the equilibrium conditions to find the reactions at the supports.

Use the definitions of bending moment and shear force to form the equations required.

(a) Left-hand reaction $= \frac{2}{3}W$. Right-hand reaction $= \frac{1}{3}W$. At x from the left-hand end if $x < L/3$ the only force to the left is the reaction.

$$\therefore V = \frac{2W}{3}; \quad M = \frac{2Wx}{3}.$$

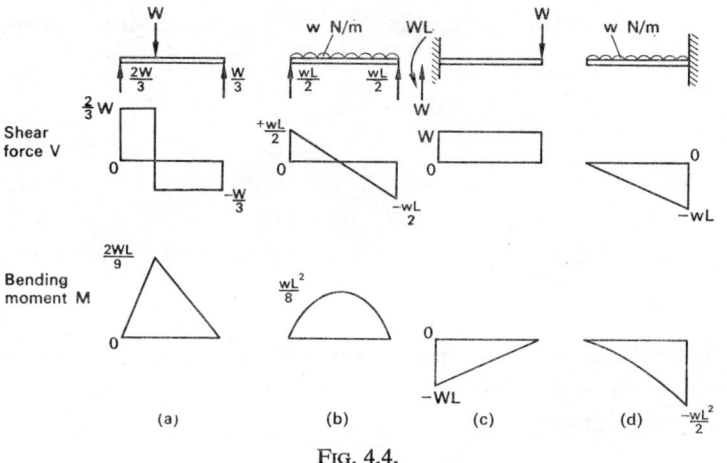

Fig. 4.4.

If $x > L/3$,

$$V = \frac{2W}{3} - W = -\frac{W}{3};$$

$$M = \frac{2Wx}{3} - W\left(x - \frac{L}{3}\right) = \frac{WL}{3} - \frac{Wx}{3}.$$

Thus V is constant with respect to x except for a change at $x = L/3$ from $+(2W)/3$ to $-W/3$.

M increases linearly with x from zero at the left-hand support to a maximum of $(2WL)/9$ at $x = L/3$ and thereafter decreases linearly to zero at $x = L$.

(b) The load is w N/m over length L m and total load is therefore wL N. Each reaction is therefore $(wL)/2$.

To the left of a section at x from the left-hand support there will be an upward (positive) force of $(wL)/2$ and a downward force due to the load of wx.

Therefore sum of forces to left of section $= V = [(wL)/2] - wx$.
At the same section the support reaction has a positive bending

moment of $[(wL)/2]x$ while the load has a negative moment of $wx(x/2)$.

$$\therefore M = \frac{wLx}{2} - \frac{wx^2}{2}.$$

Thus at $x = 0$, $V = (wL)/2$, and it decreases linearly with x to zero at $x = L/2$ and to $(-wL)/2$ at $x = L$. M is zero at each end and parabolic between with a maximum at the centre of $(+wL^2)/8$.

(c) Wall reactions are upward force W and a counter-clockwise moment WL. Along the whole length $V = W$, falling vertically to zero at the ends.

$$M = -WL + Wx.$$

(*Note:* The student who has difficulty in deciding on the sign of a bending moment component should apply a force or twist in the appropriate position and sense to a flexible ruler or piece of paper and compare the resulting shape with the sign convention given in Fig. 4.2.)

The shear-force diagram is therefore a rectangle of height W, while the bending moment diagram begins at $-WL$ and rises linearly to zero at $x = L$.

(d) For cantilevers, if one works from the free end there is no need to calculate wall reactions. The total force to the left of a section at x from the free end is wx.

$$\therefore V = -wx.$$

$$\text{Moment of this force} = -\frac{wx^2}{2}.$$

Shear-force diagram is zero at free end and falls linearly to $-wL$ at fixed end.

Bending moment diagram is zero at free end and falls parabolically to $(-wL^2)/2$ at fixed end.

The student should notice that in each of the above cases the equation for V can be obtained by differentiating with respect to x the equation for M. He is advised to master thoroughly these four simple cases since many complex problems prove to be combinations of them.

4.2. A beam ABC shown in Fig. 4.5 is simply supported at B and C, 6 m apart and has an overhang of 2 m to the end A. Point loads of 2 kN at A, 3 kN at D—which is 1 m to the right of B—and 4 kN at

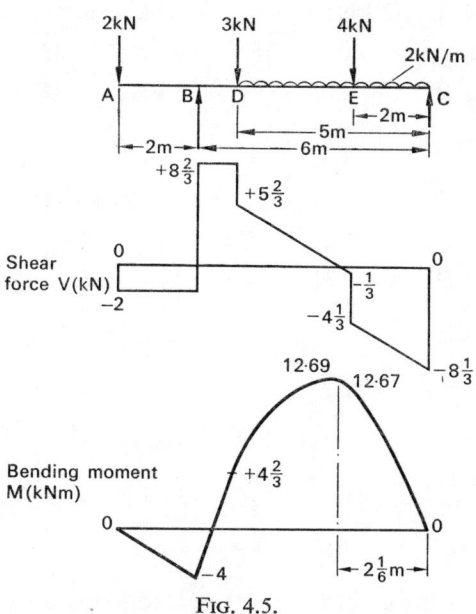

Fig. 4.5.

E, 4 m to the right of B, act vertically down. There is a uniformly distributed load of 2 kN/m over the portion from D to C.

Draw dimensioned sketches of the shearing force and bending moment diagrams.

[London Univ.]

Solution

Taking moments about C:

$$6R_B - (3 \times 5) - (4 \times 2) - (2 \times 8) - (2 \times 5 \times \tfrac{5}{2}) = 0,$$
$$R_B = 10\tfrac{2}{3} \text{ kN}.$$
$$R_B + R_C - 2 - 3 - 4 - 2 \times 5 = 0,$$
$$R_C = 8\tfrac{1}{3} \text{ kN}.$$

Over section AB:

$$V = -2 \text{ kN}; \quad M = -2x \text{ kN m}.$$
$$\text{At } A, \quad x = 0; \quad V = -2 \text{ kN}; \quad M = 0.$$
$$\text{At } B, \quad x = 2; \quad V = -2 \text{ kN}; \quad M = -4.$$

Over section BD:

$$V = -2 + 10\tfrac{2}{3} = 8\tfrac{2}{3}; \quad M = -2x + 10\tfrac{2}{3}(x-2).$$
$$\text{At } B, \quad x = 2; \quad V = 8\tfrac{2}{3}; \quad M = -4.$$
$$\text{At } D, \quad x = 3; \quad V = 8\tfrac{2}{3}; \quad M = 4\tfrac{2}{3}.$$

Over section DE:

$$V = -2 + 10\tfrac{2}{3} - 3 - 2(x-3);$$
$$M = -2x + 10\tfrac{2}{3}(x-2) - 3(x-3) - 2\frac{(x-3)^2}{2}.$$
$$\text{At } D, \quad x = 3; \quad V = 5\tfrac{2}{3}, \quad M = 4\tfrac{2}{3}.$$
$$\text{At } E, \quad x = 6; \quad V = -\tfrac{1}{3}; \quad M = 12\tfrac{2}{3}.$$

In order to draw the diagrams one needs to know the point along the section where $V = 0$. But V falls linearly from

$$5\tfrac{2}{3} \text{ at } x = 3 \text{ to } -\tfrac{1}{3} \text{ at } x = 6,$$

that is, it is zero at a point which divides the section of beam into the ratio $5\tfrac{2}{3} : \tfrac{1}{3}$, i.e. the point is at $2\tfrac{5}{6}$ m from D. Here M is maximum and equal to 12·69 kN m.

Over section EC:

It is easier to work from the right since the summation has fewer terms, at z from point C:

$$V = -8\tfrac{1}{3}+2z; \quad M = +8\tfrac{1}{3}z - 2(z^2/2).$$
At E, $z = 2$; $V = -4\tfrac{1}{3}$; $M = +12\tfrac{2}{3}$.
At C, $z = 0$; $V = -8\tfrac{1}{3}$; $M = 0$.

From examination of the equations and values for the various sections of the beam the diagrams in Fig. 4.5 can be drawn.

4.3. A horizontal beam of length 10 m is simply supported at the ends and carries a uniformly distributed load of 30 kN/m along its whole length. Counter-clockwise moments are applied to the beam, 120 kN m at the left-hand support and 100 kN m at the right-hand support.

Draw, approximately to scale, the bending-moment diagram for the beam and find the point of maximum bending moment and the point of contraflexure (if any). [London Univ.]

Solution

The loading diagram is shown in Fig. 4.6.

Taking moments about the left-hand support with counter-clockwise moments positive,

$$120 - 30 \times 10 \times 5 + 100 + R_R \times 10 = 0,$$
$$R_R = -12 + 150 - 10$$
$$= 150 - 22 = 128 \text{ kN}.$$

For equilibrium sum of vertical forces = 0,

$$R_L - 30 \times 10 + R_R = 0,$$
$$R_L = 300 - 128,$$
$$R_L = 172 \text{ kN}.$$

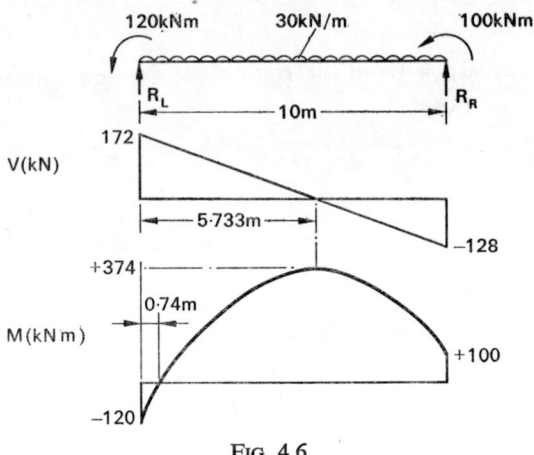

Fig. 4.6.

Shear force at x metres from the left-hand end is the sum of the forces to the left of the section:

$$V = (172 - 30x) \text{ kN.}$$

This is true for the whole length of the beam

when $\qquad V = 0$
i.e. $\qquad 172 - 30x = 0,$
$$x = \tfrac{172}{30}$$
$$= 5\tfrac{11}{15}$$
$$= 5\cdot733 \text{ m.}$$

Bending moment at x metres from left-hand end is the sum of the moments to the left of the section:

$$M = -120 + 172x - 30\frac{x^2}{2}.$$

Maximum value is at $x = 5\cdot733$,

$$M = -120 + 986 - 492 = 374 \text{ kN m.}$$

Point of contraflexure occurs where the bending moment passes through zero (this point is also called "point of inflection"),

i.e.
$$-120 + 172x - 15x^2 = 0,$$
$$x^2 - 11\cdot 48x + 8 = 0,$$
$$x = 5\cdot 74 \pm \sqrt{(5\cdot 74^2 - 8)},$$
$$x = 5\cdot 74 \pm \sqrt{(32\cdot 95 - 8)},$$
$$x = 0\cdot 74 \text{ m}.$$

Hence the diagrams in Fig. 4.6 can be drawn.

Note: This method of writing an equation for V or M and equating it to zero will always locate the maximum bending moment or point of contraflexure. If however, there are point loads as well as a distributed load on the beam, the diagrams are discontinuous and no one equation applies over the whole beam. The procedure recommended in such cases is to draw the shear-force and bending-moment diagrams in order to find the section of the beam where the required points lie and then to write the equations for those sections and equate to zero. It is seldom worth the trouble of drawing diagrams to scale and measuring off the locations of the required points.

4.4. Find the second moment of area about the neutral axis of the beam cross-section shown in Fig. 4.7.

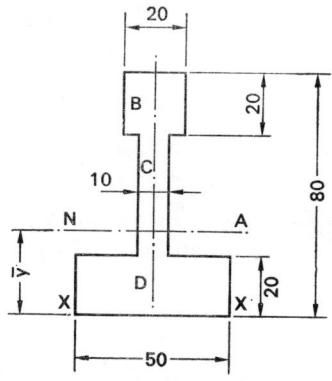

All dimensions in mm

Fig. 4.7.

Solution

The method which follows is a "sausage-machine" method which will deal with all types of sections. For some symmetrical sections other methods are faster, but even there the method below has the merit that all numerical calculations are displayed so that checking is easy.

A table such as Table 4.1 is drawn up in which all measurements are in centimetres.

TABLE 4.1

Part	Area	Distance to centroid (y) from XX	Ay	I_{CG}	$(y-\bar{y})$	$(y-\bar{y})^2$	$A(y-\bar{y})^2$
B	4	7	28	$1\frac{1}{3}$	4	16	64
C	4	4	16	$5\frac{1}{3}$	1	1	4
D	10	1	10	$3\frac{1}{3}$	-2	4	40

$$\Sigma A = 18; \quad \Sigma Ay = 54; \quad \Sigma I_{CG} = 10; \quad \Sigma A(y-\bar{y})^2 = 108.$$

$$\bar{y} = \frac{\Sigma Ay}{\Sigma A} = \frac{54}{18} = 3; \quad I_{NA} = I_{CG} + A(y-\bar{y})^2$$

$$= 10 + 108$$

$$= 118 \text{ cm}^4.$$

$$\therefore I_{NA} = 1 \cdot 18 \times 10^{-6} \text{ m}^4.$$

The section is divided into parts whose areas A are readily calculated and whose centroid distances y from a datum at the base of the section can be determined by inspection. The distance from the datum to the centroid of the whole figure $\bar{y}$ will be

$$\bar{y} = \frac{\Sigma Ay}{\Sigma A}.$$

The second moment of each part about an axis parallel to the datum passing through its own centroid is then calculated (I_{CG}). [If the parts have been chosen to be rectangles, $I_{CG} = bd^3/12$.]

In order to use the parallel axes theorem, $A(y-\bar{y})^2$ is then calculated and the second moment about the centroid of the cross-section can be obtained by summation. Since the neutral axis passes through the centroid, this will be the required value of I_{NA}.

Calculations have been carried out in centimetre units for convenience, but the final answer quoted in (metres)⁴ since this is the appropriate S.I. unit. It should be noted that since $1 \text{ cm} = 10^{-2}$ m,
$$1 \text{ cm}^4 = (10^{-2})^4 \text{ m}^4$$
$$= 10^{-8} \text{ m}^4.$$

4.5. An I-section girder 300 mm deep has its web 25 mm thick while the top flange is 150 mm wide and 25 mm deep and the bottom flange is 200 mm wide by 25 mm deep. If the girder is simply supported at its ends, find the maximum span which can be used if the total distributed load per metre run is 6 kN and the maximum stress in the beam material is limited to 70 MN/m². [London Univ.]

Solution

By the method of the previous example, the neutral axis is located and found to be at 138·6 mm above the bottom flange and I_{NA} is found to be 1.97×10^{-4} m⁴.

Stress f due to bending at a distance y from the neutral axis is given by the bending stress equation, i.e. $f = My/I$.

In a simply supported beam with uniformly distributed load, the maximum bending moment is $(wL^2)/8$.

Stress will be a maximum at the extreme fibre of the beam, i.e. at the top of the top flange where $y = 0.3 - 0.1386$
$$= 0.1614 \text{ m}.$$

f is limited to 70×10^6 N/m².

$$\therefore M = \frac{fI}{y}$$

$$\frac{wL^2}{8} = \frac{70 \times 10^6 \times 1.97 \times 10^{-4}}{1.614 \times 10^{-1}},$$

$$L^2 = \frac{70 \times 1\cdot 97 \times 8 \times 10^2}{6 \times 10^3 \times 1\cdot 614 \times 10^{-1}}$$

$$= 114 \text{ m}^2.$$

$$L = 10\cdot 7 \text{ m}.$$

4.6. A symmetrical I-section beam has the dimensions shown in Fig. 4.8. At a certain cross-section the bending moment is found to be

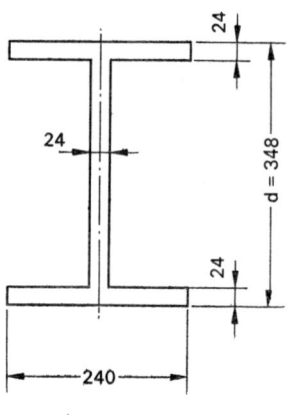

All dimensions in mm

FIG. 4.8.

82 kN m and the shear force is 130 kN. Sketch the distribution of bending stress and shear stress across the centre line of a cross-section and show values at key points.

Solution

Since the section is symmetrical about a horizontal centre line, the centroid and neutral axis must be at mid-depth.

By calculation (see example 4.4), $I_{NA} = 35\cdot 66 \times 10^{-5}$ m⁴.

Bending stress f at y from the neutral axis is $f = My/I$, i.e. $f \propto y$.

At the outside of the beam $y = 0.174$ m.

$$\therefore f = \frac{82 \times 10^3 \times 0.174}{35.66 \times 10^{-5}}$$

$$= \frac{820 \times 1.74}{35.66} \times 10^6$$

$$= 40 \text{ MN/m}^2.$$

Shear stress q at y from the neutral axis, if the breadth at this distance is b, is

$$q = \frac{V}{Ib} \int_{y_1}^{y_2} y \, dA.$$

Consider an element of area in the top flange; there $dA = b_f \, dy$,

$$q = \frac{V}{Ib} \int_{y_1}^{d/2} b_f y \, dy$$

$$= \frac{V b_f}{Ib} \left[\frac{y^2}{2} \right]_{y_1}^{d/2}$$

$$= \frac{V b_f}{2Ib} \left[\left(\frac{d}{2} \right)^2 - y_1^2 \right].$$

In the flange, the breadth is uniform, and above the bottom layer $b = b_f$.

$$\therefore q = \frac{V}{2I} \left[\frac{d^2}{4} - y_1^2 \right].$$

To find the stress at the top of the flange, put $y_1 = d/2$, and one finds $q = 0$.

This clearly must be so, for there can be no shear stress at a free surface which is not carrying an applied shear force. The shear stress then increases parabolically until, while still in the flange, $y_1 = 0.150$ m.

Here
$$q = \frac{130 \times 10^3}{2 \times 35.66 \times 10^{-5}} \cdot [0.174^2 - 0.150^2]$$

$$= 1.4 \text{ MN/m}^2.$$

Moving infinitesimally lower, one passes from flange into web; here the breadth at the level of the shear stress changes from the breadth of flange b_f to the breadth of the web b_w.

Thus
$$q = \frac{Vb_f}{2Ib_w}\left[\frac{d^2}{4} - y_1^2\right]$$

$$= 14 \text{ MN/m}^2,$$

i.e. at the point where the flange joins the web the shear stress increases from 1·4 MN/m² to 14 MN/m²,

i.e.
$$\left[\frac{b_f}{b_w} \times 1\cdot 4\right].$$

In the remainder of the web one needs to evaluate

$$\int_{y_1}^{d/2} y\, dA,$$

since there is a discontinuity due to the change in breadth.

Now the integral is the definition of the first moment of area, about the neutral axis, of the portion of cross-section above the level y_1; it is therefore equal to $\Sigma\, A\bar{y}$, where A represents an area whose centroid can be located by inspection and $\bar{y}$ is the distance to the centroid from the neutral axis.

Therefore if y_1 is anywhere in the web,

$\Sigma A\bar{y}$ = area of flange × distance to centroid of flange + area of web above y_1 × distance to centroid of this area.

$$\Sigma A\bar{y} = 5\cdot 76 \times 10^{-3} \times 0\cdot 162 + 0\cdot 024(0\cdot 150 - y_1)\frac{(0\cdot 150 + y_1)}{2}.$$

$$\therefore q = \frac{V}{Ib_w}[93\cdot 2 \times 10^{-5} + 0\cdot 012(0\cdot 150^2 - y_1^2)].$$

At $y_1 = 0\cdot 15$, i.e. at top of web, $q = 14$ MN/m² as before. Below this there is a parabolic distribution until $y_1 = 0$ (i.e. at the neutral axis),

$$q = \frac{130 \times 10^{-3}}{35 \cdot 66 \times 10^{-5} \times 0 \cdot 024} (93 \cdot 2 \times 10^{-5} + 0 \cdot 012 \times 0 \cdot 150^2)$$

$$= 18 \cdot 3 \text{ MN/m}^2,$$

and the form of the bending and shear-stress distributions are as shown in Fig. 4.9.

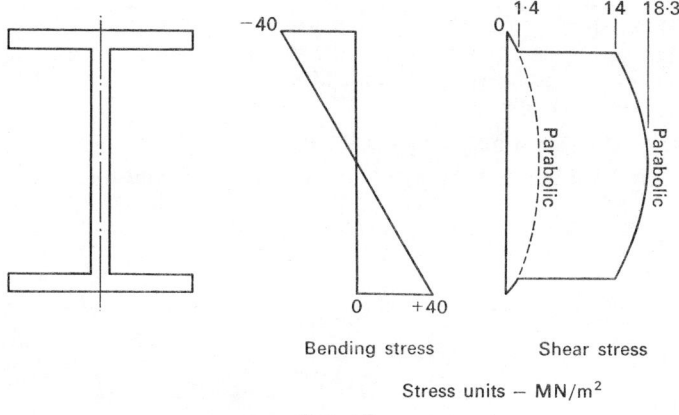

Bending stress   Shear stress

Stress units — MN/m²

FIG. 4.9.

4.7. A short length of tube having an outside diameter of 100 mm and an inside diameter of 75 mm is subjected to a longitudinal compressive force of 200 kN acting along a line parallel to the axis of the tube. If the maximum compressive stress produced is 90 MN/m², find (a) the minimum stress, (b) the eccentricity of the compressive force, and (c) the greatest eccentricity the force could have without causing a reversal in stress. [London Univ.]

Solution

Referring to Fig. 4.10 it is clear that the effect of a load P at an eccentric distance e will be the same as the effect of an axial force of the same magnitude P together with a couple of magnitude Pe. The

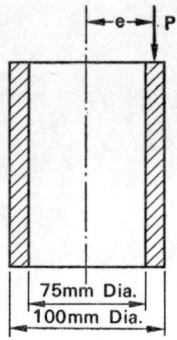
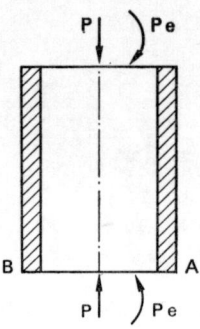

FIG. 4.10.

stresses due to each may be calculated separately and, by the principle of superposition, the total stress obtained by summation, i.e.

$$f = \frac{P}{A} + \frac{My}{I},$$

where P/A is the direct stress and My/I is the bending stress.

Here
$$\frac{P}{A} = \frac{2 \times 10^5}{\pi/4(0 \cdot 1^2 - 0 \cdot 075^2)} = \frac{2 \times 10^5}{\pi/4(0 \cdot 175 \times 0 \cdot 025)}$$
$$= 58 \cdot 2 \text{ MN/m}^2.$$

$$I = \frac{\pi}{64}(D^4 - d^4),$$

$$I = \frac{\pi}{64}(D^2 + d^2)(D^2 - d^2) = 3 \cdot 355 \times 10^{-6} \text{ m}^4.$$

$M = 200\,000 e$ N m.

(a) The minimum stress will be at the point where the maximum tension due to bending is present to offset the compressive stress, i.e. at point B (Fig. 4.10) the maximum stress will be at point A where both compressive stresses combine. For both points, $y = 0 \cdot 05$ m.

$$\frac{My}{I} = \frac{200\,000 e \times 0 \cdot 05}{3 \cdot 355 \times 10^{-6}}.$$

But bending stress = maximum stress − compressive stress

$$= 90 \times 10^6 - 58 \cdot 2 \times 10^6$$
$$= 31 \cdot 8 \times 10^6 \text{ N/m}^2.$$
$$\therefore e = \frac{31 \cdot 8 \times 10^6 \times 3 \cdot 355 \times 10^{-3}}{2 \times 10^5 \times 5 \times 10^{-2}}$$
$$= 10 \cdot 67 \text{ mm}.$$

(b) Minimum stress = compressive stress + maximum tension due to bending

$$= -58 \cdot 2 + 31 \cdot 8$$
$$= -26 \cdot 4 \text{ MN/m}^2.$$

(c) For stress reversal the eccentricity e must be sufficient for the bending tensile stress to exceed the direct compressive stress. At the critical value of e both are equal.

$$\therefore \frac{P}{A} = \frac{Pey}{I},$$
$$e = \frac{I}{Ay}.$$
$$\therefore e = 19 \cdot 5 \text{ mm}.$$

Problems

4.8. Sketch the bending-moment and shear-force diagrams for:

(a) A beam length L simply supported at the ends with equal and opposite couples C applied to the beam at the supports—the couple at the left-hand support is clockwise.

(b) A beam length L pin-supported at the ends with a clockwise couple C, acting on the beam at the left-hand support only.

(c) A uniform wooden beam length L floating level in water with a vertical point load P acting down at the mid-point (neglect weight of beam).

[(a) $V = 0$ over whole length; $M = +C$ over whole length.
(b) $V = -C/L$ over whole length; M falls linearly from $+C$ at the left-hand end to zero at the right.
(c) V rises linearly from 0 at left-hand end to $+P/2$ at centre; falls to $-P/2$ and then rises linearly to 0 at the right-hand end. M is zero at both ends and rises in a parabola (concave upwards) to $PL/8$ at the centre.]

4.9. A horizontal cantilever 5 m long carries a uniformly distributed load of 30 kN/m over the whole length and a point load of 50 kN 1 m from the free end. Draw the bending-moment and shear-force diagrams and find the maximum values of each.

[$V_{max} = 200$ kN at the wall. $M_{max} = -575$ kN m at the wall.]

4.10. A horizontal pin-supported beam is 5 m long; a clockwise couple of 10 000 N m is applied to the beam at the left-hand support and a counter-clockwise couple of 50 000 N m is applied at the right-hand support. Find (a) the reaction at each support, and (b) the maximum shear force and bending moment.

[$R_L = 8000$ N upwards; $R_R = 8000$ N downwards. $V_{max} = 8000$ N and it is constant all along the beam. $M_{max} = 50\,000$ N m at the right-hand support.]

4.11. The same beam as problem 4.10 has, instead of the two couples at the supports, a single clockwise couple of 20 kN m acting at mid-span. Find the reactions at each support and the location and magnitudes of maximum shear force and bending moment.

[$R_L = 4$ kN downwards; $R_R = 4$ kN upwards. $V_{max} = -4000$ N on the whole length. $M_{max} = +10$ kN m at mid-span.]

4.12. A uniform horizontal beam 8 m long carries a uniformly distributed load, including its own weight, of 20 kN/m and, in addition, a concentrated load of 20 kN at the left-hand end. There are two simple supports 5 m apart and positioned such that each support carries half the total load. Draw shear-force and bending-moment diagrams and show the principal values on each of them.

[Left-hand support at 1·05 m from l.h. end of beam. Shear force values are: l.h. end, − 20; l.h. support − 41 and +49; r.h. support − 51 and +39; zero at 3·5 m from l.h. end (units kN). Bending-moment values are: l.h. support − 32; maximum +28; r.h. support − 38 (units kN m).]

4.13. A uniform horizontal beam carries a uniformly distributed load of 5 kN/m and has a counter-clockwise moment of 50 kN m at the left-hand simple support and a clockwise moment of 75 kN m at the right-hand simple support. The length is 10 m and the supports are at the ends. Calculate the position of the point of inflection and draw the shear-force and bending-moment diagrams showing the values at key points.

[Points of inflection at 4 m and 5 m from l.h. support. Shear force: 22·5 kN at l.h. end; 0 at 4·5 m from l.h. end; − 27·5 at r.h. end. Bending moment: − 22·5 kN m at l.h. end; 0 at 4 m and 5 m, 0·625 (positive maximum) at 4·5 m from l.h. end; − 27·5 at r.h. end.]

4.14. Calculate the shear force and bending moment at a point 1 m to the right of the left-hand support of a uniform beam simply supported at the ends over a span of 3 m if the loading is uniformly distributed loads of 6 kN/m over 1 m at each end, a point load of 2 kN at 1 m from the left-hand support, and a clockwise moment of 3 kN m at the mid-point.

[$V = +333$ N. $M = +3333$ N m.]

4.15. A horizontal beam AB, 9 m long, carries a uniformly distributed load over the whole length which totals 450 kN. The beam is supported at the end A and at a point C distant d metres from the end B. What must be the value of d if the mid-point of the beam is to be a point of inflection? For this arrangement, draw the shear-force and bending-moment diagrams showing the principal values on each.

[$d = 3$ m. Shear force: 112·5 kN at l.h. end; 0 at 2·25 m to the right; $-187·5$ kN and $+150$ kN at r.h. support. Bending moment: $+126·5$ kN m at 2·25 m to the right of l.h. support; 0 at mid-point; -225 kN m at r.h. support.]

4.16. A horizontal cantilever 6 m long has a frictionless hinge at 1 m from the built-in end; it has a vertical prop at the free end and loads of 10 kN vertically downwards at 1 m and 2 m from the free end. Draw shear-force and bending-moment diagrams for the whole beam.

[Shear force: $+14$ kN at prop; $+14$ kN and $+4$ kN at first load; $+4$ kN and -6 kN at second load; -6 kN from second load to wall. Bending moment: $+14$ kN m at first load; $+18$ kN m at second load; 0 at hinge; -6 kN m at wall.]

(*Note.* Since the hinge is frictionless it cannot transmit any moment, only force. This is the same as the situation at the centre pin of a three-pinned arch analysed in Chapter 1.)

4.17. A horizontal beam *ABCDEF* has three supports, *A* at the left-hand end, *C* 4 m from *A*, and *F* at the right-hand end, 8 m from *A*. Point loads of 5 kN act downwards at *B*, 2 m from *A*, and *E* 1 m from *F*. There is a frictionless hinge at *D* midway between *C* and *F*. Find the values of shear force in each section and of bending moment at each of the lettered points along the beam.

[Shear force: $AB+1·25$ kN; $BC-3·75$ kN; $CD+2·5$ kN; $EF-2·5$ kN. Bending moment: A 0; $B+2·5$ kN m; $C-5$; D 0 (hinge); $E+2·5$; F 0.]

4.18. A friend of yours is thinking of fitting a beam in his house so that he can remove a dividing wall between two downstairs rooms. He has the choice of three beams, all of the same material and weight per unit length and he wishes to use the one which will carry the greatest uniformly distributed load without the compressive stress exceeding the safe value for the material. The beams have the following cross-sections: (a) square 100 by 100 mm, (b) circle, diameter $200/\sqrt{\pi}$ mm, (c) I-section 125 mm deep, 100 mm wide; top flange 25 mm, bottom flange 50 mm, and web 50 mm thick.

Explain which beam will be best and why; calculate the ratio of the maximum bending moments that each can carry.

[Ratio of moments: circle : square : I-section $= 1 : 1\cdot19 : 1\cdot68$.]

4.19. Equal and opposite couples are applied at each end of a thin, steel rule 20 mm wide by 0·5 mm thick. The rule bends so that it lies along the arc of a circle; it subtends an angle of 60° at the centre and the arc length is 200 mm. ($E = 200$ GN/m².)

Calculate (a) the bending moment in the rule, (b) the maximum bending stress in the rule, and (c) sketch the shear-force and bending-moment diagrams for the system.

[(a) 0·218 Nm. (b) 262 MN/m². (c) Shear-force diagram is zero throughout. Bending-moment diagram 0·218 throughout.]

4.20. Find the required distance between centres of timber floor joists 180 mm deep, 76 mm wide spanning 4·6 m if the load carried is to weigh 2·5 kN/m² of floor area (this includes the weight of the joists themselves). The bending stress in the timber must not exceed 5 MN/m². [London Univ., 1966]

[310 mm. (N.B.—In this type of problem the assumption is made that each joist is responsible for supporting all the load on the strip of floor which extends $d/2$ on either side of the joist centre line, where d is the distance between joist centres. The problem then reduces to that of a beam with a distributed load.)]

4.21. Repeat problem 4.20 with a loading of 16 kN/m²; maximum permissible stress 7 MN/m²; joist dimensions: width 115 mm, depth 300 mm, span 4·25 m.

[334 mm.]

4.22. A hollow steel pipe, outside diameter 100 mm, bore 80 mm, is simply supported over a span of 2·5 m and carries a single point load at mid-span. Calculate the maximum value of the load if the stress is limited to 120 MN/m². What will be the radius of curvature of the centre of the beam if $E = 200$ GN/m²?

[11·14 kN. 83·33 m.]

4.23. A tapered shaft of length L is built-in at the larger end (diameter D) and is free at the smaller end (diameter d). A force W is applied at the free end perpendicular to the shaft axis. Show that the maximum bending stress at a section distance x from the free end of the shaft is

$$\frac{Wx}{\frac{\pi}{32}\left[d+(D-d)\frac{x}{L}\right]^3}.$$

Hence find the value of x at which the greatest value of bending stress occurs. [London Univ.]

$$\left[x = \frac{dL}{2(D-d)}\right]$$

4.24. A steel tube 42 mm outside diameter and 32 mm inside diameter is used as a simply supported beam on a span of 2 m when it is found that the maximum safe load it can carry at mid-span is 1·25 kN.

Four of the tubes are then placed together to form a single beam with the centres of the tubes at the corners of a square of 42 mm side. If the tubes are firmly fixed together and supported so that one pair of tubes has centres immediately above the centres of the other pair, what is the maximum central load the composite beam can carry if the span and the maximum stress in the material are the same as for the single tube beam? [London Univ.]

[8·85 kN.]

4.25. A flagstaff is fixed vertically into the ground and is 9 m tall. It is of square cross-section tapering from 150 mm square at the bottom to 75 mm square at the top. A horizontal force of 310 N is applied at the top acting in the direction of one of the diagonals of the square. Calculate the maximum stress due to bending.
[London Univ.]

[83·12 MN/m².]

4.26. A beam of section shown in Fig. 4.11 is 3·2 m long and simply supported at each end. At 1·22 m from each support there is a point load of magnitude W. Find the maximum value of W if the maximum

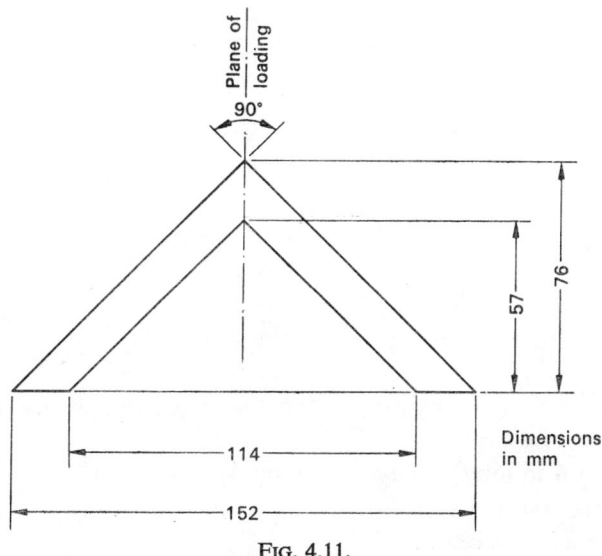

Fig. 4.11.

tensile stress in the material must not exceed 125 MN/m² and the maximum compressive stress must not exceed 110 MN/m². With this value of W, what are the maximum tensile and compressive stresses and where along the beam do they occur? [London Univ.]

[$W = 2096$ N. $f_{c,\max} = 110$ MN/m², $f_{T,\max} = 85·4$ MN/m² throughout the length of beam between the two loads.]

4.27. A light alloy beam is of the section shown in Fig. 4.12. Find (a) the position of the neutral axis NA, (b) I_{NA}, and (c) the maximum central point load the beam will carry when simply supported on a span of 0·75 m and the maximum allowable bending stress in both tension and compression is 125 MN/m². [London Univ.]

[(a) $\bar{y} = 13\cdot3$ mm. (b) $6\cdot258 \times 10^{-8}$ m⁴. (c) 2·24 kN.]

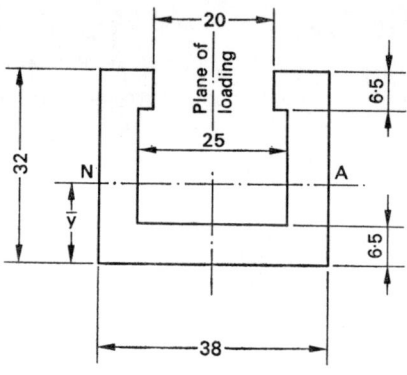

All dimensions in mm

Fig. 4.12.

4.28. Deduce an expression for the shearing stress at any point on the section of a beam.

A beam 6 m long, having a rectangular section 0·3 m by 0·225 m is subjected to a point load of 10 kN at mid-span. Find the maximum shear stress in the beam.

[$11\cdot1 \times 10^4$ N/m².]

4.29. Find the ratio of the maximum to the mean shearing stress produced by a shearing force on the section of (a) a rectangular beam, and (b) a circular beam.

[(a) 3 : 2. (b) 4 : 3.]

STRESSES IN BEAMS

4.30. Show that the distribution of shearing stress across the vertical section of a loaded beam of rectangular cross-section is parabolic. Find the maximum shearing stress produced by a load of 200 kN in the vertical section of a hollow beam of square section if the outside width is 125 mm and the thickness of material is 25 mm.

[London Univ.]

[43·23 MN/m².]

4.31. A symmetrical rolled steel joist of I-section is used as a beam with the web vertical. The section is 125 mm deep and 62·5 mm wide; the web is 5 mm thick and the flanges 8·75 mm. The beam carries a shear load of 50 kN.

Draw to scale a diagram showing the distribution of shear stress in the beam.

[Maximum shear stress 92·47 MN/m². Shear stress at junction of web and flange: 6·03 MN/m², 75·36 MN/m².]

4.32. Two beams, particulars of which are given in Table 4.2, are simply supported at the ends over equal spans and carry central loads to give the same maximum stress. Determine the ratio of the maximum shear stress in the webs. [London Univ.]

TABLE 4.2
(Dimensions in mm)

Section	Web thickness	Flange thickness	Total flange width	Total depth	Distance of neutral axis from outer edge of flange
I	5	8·75	62·5	125	62·5
T	12·5	12·5	125	100	26·9

[$q_I/q_T = 3·42$.]

4.33. A rod of circular section is subjected to a shearing force on a plane perpendicular to its axis. Find the maximum shearing stress in terms of the shearing force and rod diameter.

If the rod is used as a beam with free ends and a central concentrated load, express the free length in terms of the diameter for which the maximum shearing stress, due to the shearing force, is half the maximum direct stress.

$$\left[q_m = \frac{16V}{3\pi D^2}. \quad L = \frac{2}{3}D.\right]$$

4.34. A steel bar of rectangular section 80 mm by 20 mm is simply supported at its ends over a span of 2 m with its longer side vertical. What is the maximum stress due to bending if the bar supports a uniformly distributed load of 1 kN/m? If at the same time a tensile force of 10 kN acts along the longitudinal centre line of the bar, what will then be the maximum tensile and compressive stresses?

[± 23.5 MN/m². $+29.75$ MN/m². $+17.25$ MN/m².]

4.35. A short column of external diameter D, internal diameter d, carries an eccentric load P; find an expression for the greatest eccentricity which the load can have without producing tension in the cross-section of the column.

A cast-iron column, external diameter 150 mm, internal diameter 115 mm, carries a vertical compressive load of 200 kN. Find the maximum eccentricity of the load if the tensile stress in the column is not to exceed 30 MN/m². What will then be the value of the maximum compressive stress? [London Univ.]

$$\left[\frac{(D^2+d^2)}{8D}. \quad 62.5 \text{ mm.} \quad 84.9 \text{ MN/m}^2.\right]$$

4.36. A 50 mm by 12·5 mm flat steel bar was placed in a testing machine and subjected to a tensile load of 60 kN parallel to the centre line but offset 12·5 mm from the centre line of the 50 mm face. An extensometer placed in line with the load recorded an extension of 168 μm on a gauge length of 200 mm. Calculate the maximum and minimum stresses set up and the value of Young's modulus.

[London Univ.]

[240 MN/m² tensile. 48 MN/m² compressive. $E = 200$ GN/m².]

4.37. A steel plate 100 mm wide and 9·5 mm thick is pin-jointed at the ends and subjected to a tensile force of 50 kN acting parallel with the bar axis but displaced 12·5 mm from the axis along the centre line of the width. The distance between the centres of the pin joints is 2·125 m. What load at the centre of the plate acting perpendicular to the axis of the plate will cause a uniform distribution of longitudinal stress? [London Univ.]

[1·175 kN.]

4.38. A bar of rectangular cross-section 6·25 mm thick and 75 mm wide has a rectangular recess 25 mm long, 12·5 mm deep milled across the thickness (Fig. 4.13). A tensile load of 40 kN acts along the axis

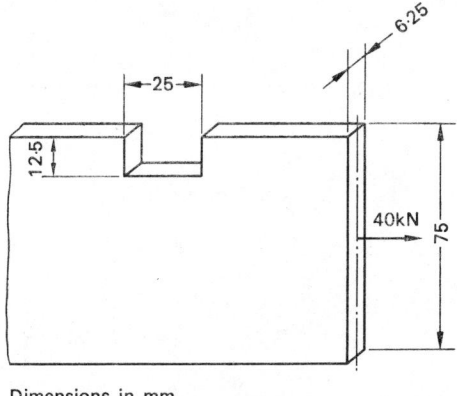

Dimensions in mm

FIG. 4.13.

of the original bar; find the stress distribution across the portion of the bar midway along the recess and calculate the greatest depth of recess which may be cut if the maximum stress beneath the recess is not to exceed 185 MN/m². (Neglect local end effects of the recess.) [London Univ.]

[Stress varies linearly from 163·8 MN/m² tensile at base of recess to 41 MN/m² tensile at surface of bar. Maximum depth = 15·3 mm.]

4.39. A column of square cross-section 150 mm side carries an axial compressive load of 300 kN. A hole of 100 mm diameter is then bored down the column so that its centre lies on one centre line of the cross-section but is 15 mm from the other centre line (Fig. 4.14). Find the stresses at points C and D.

[At C 26·3 MN/m² compressive. D 15·77 MN/m² compressive.]

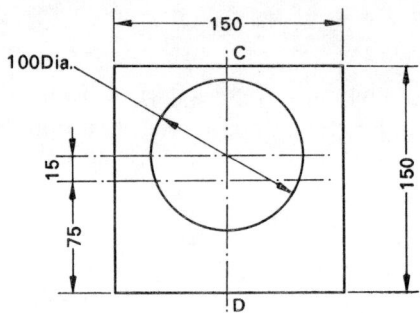

All dimensions in mm
Fig. 4.14.

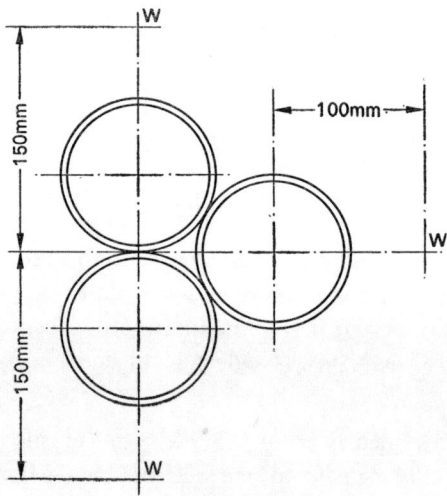

Fig. 4.15.

4.40. Three tubes are welded together to form a short column. Both ends are covered with rigid plates and three equal compressive loads are applied at the points W (Fig. 4.15). The tube's outside diameter is 0·1 m and the bore 0·09 m. Find the magnitude of the applied loads if the maximum compressive stress in the combined column is 100 MN/m².

[65·25 kN.]

CHAPTER 5

Torsion

Definitions and Theory

(a) When a torque T (Nm) is applied to a parallel shaft of circular cross-section whose polar second moment of area is J (m⁴), the shear stress q (N/m²) at radius r (m) is given by

$$\frac{q}{r} = \frac{T}{J}.$$

Also, if two cross-sectional planes a distance L (m) apart twist relative to one another through an angle θ (rad), the quantities mentioned are all connected by the general torsion equation

$$\frac{T}{J} = \frac{q}{r} = \frac{G\theta}{L},$$

where G (N/m²) is the modulus of rigidity of the shaft material.

N.B.—In form this relation is similar to the simple bending equation

$$\frac{M}{I} = \frac{f}{y} = \frac{E}{R}.$$

(b) The **assumptions** made in deriving this relation are:

(1) The shaft is straight and of uniform cross-section.
(2) The torque is constant over the length L.
(3) Cross-sections which were planes before twisting remain so after twisting.
(4) Lines which were radial before twisting remain radial after twisting.
(5) Induced stresses do not exceed the limit of proportionality.

TORSION

(c) If the shaft is rotating at ω rad/s and carrying a torque T Nm, the power P watts transmitted, is

$$P = T\omega.$$

(d) If two shafts are joined end to end by a flanged coupling so that torque T is transmitted across the joint (Fig. 5.1), the shear stress q_K in a key holding a flange to the shaft must be such that $q_K A_K D/2 = T$, where A_K is the area carrying the shear stress and $D/2$ is the radial distance from the shearing surface to the shaft axis.

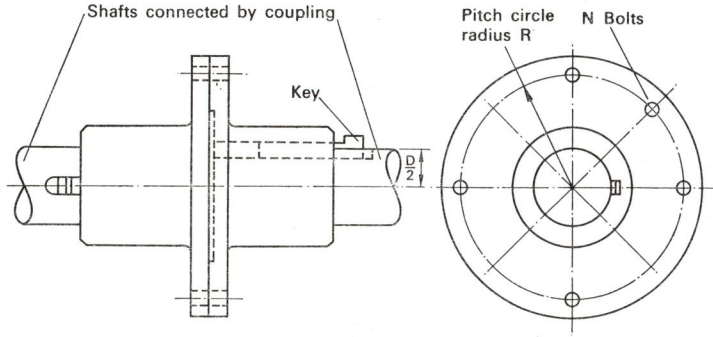

FIG. 5.1. Flanged coupling.

Similarly, if N pins or bolts each of cross-sectional area A_B are arranged on a circle of radius R, the shear stress in the bolts q_B is given by $q_B A_B N R = T$ assuming that the shear stress is uniformly distributed across the bolt cross-section.

(e) The shear stresses given by the equations in paragraph (a) act in a circumferential direction, and therefore an elemental square drawn about a point at radius r (Fig. 5.2) would have shear stresses q acting on each cross-sectional face. From Chapter 3 it is known that complementary shear stresses act on the longitudinal faces of the element and the two together give rise to tension and compression on oblique planes. Thus a shaft carrying pure torque may fail in tension or compression on an oblique plane.

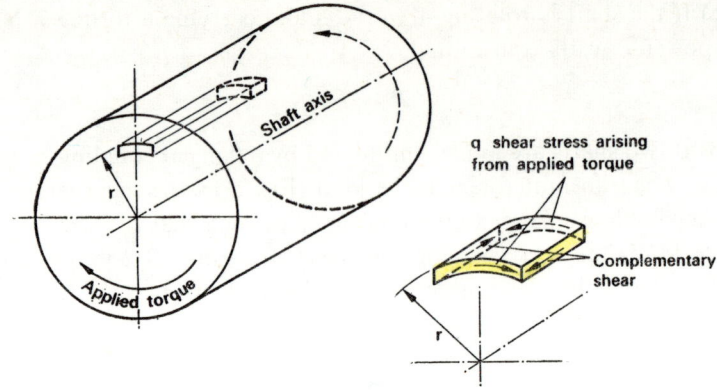

FIG. 5.2. Shear stresses at a point at distance *r* from the axis of a shaft carrying a torque.

(f) If a shaft is undergoing positive bending at the same time as it is transmitting a torque, the stress system at any point between the centre and the upper surface of the shaft will be as shown in Fig. 5.3a;

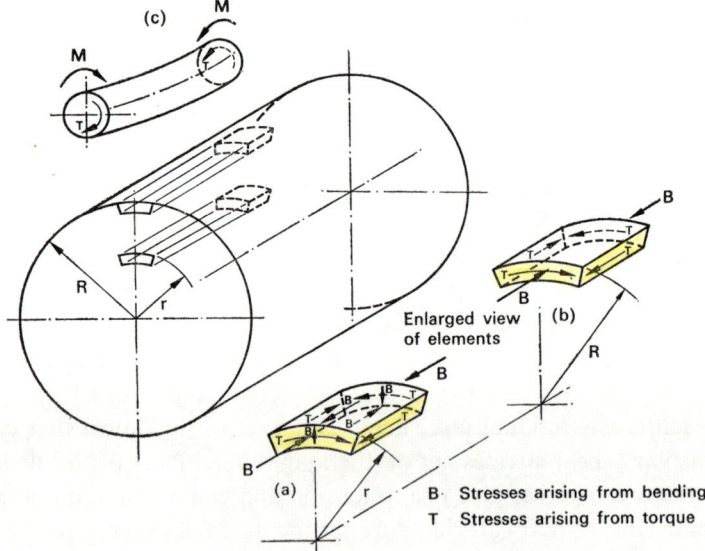

FIG. 5.3. Combined torsion and bending.

TORSION

at the surface the shear stress due to bending is zero and the stress system simplifies to that shown in Fig. 5.3b. In order to find the magnitude and direction of resulting direct and shear stresses the methods of Chapter 3 can be used.

Worked Examples

5.1. A small motor-boat has a bronze propeller shaft which turns at 160 rad/s and transmits a power of 45 kW. If the shear stress due to the torque is not to exceed 14 MN/m², find (a) a suitable shaft diameter, (b) the angle of twist in radians over a length of 3 m. ($G = 40$ GN/m².)

Solution

(a) The stress must be below the elastic limit and the formula $\frac{T}{J} = \frac{q}{r} = \frac{G\theta}{L}$ therefore applies.

Also power (P) = torque $(T) \times$ angular velocity (ω)

$$T = P/\omega = 45\,000/160 \text{ Nm.}$$

J for a solid circular section $= \pi D^4/32$.
q will be maximum at the shaft surface where $r = D/2$.

$$\therefore \frac{45\,000}{160} \times \frac{32}{\pi D^4} = \frac{14 \times 10^6}{D/2}$$

$$D^3 = \frac{45 \times 10^3 \times 32}{160\pi \times 14 \times 10^6 \times 2}$$

$$= \frac{4 \cdot 5 \times 3 \cdot 2 \times 10^{-4}}{1 \cdot 6\pi \times 2 \cdot 8},$$

$$D = 0 \cdot 04677 \text{ m,}$$

$$D = 46 \cdot 77 \text{ mm,}$$

i.e. shaft diameter must be at least 46·77 mm. In practice the nearest available size above this would be used.

(b) Angle of twist

$$\theta = \frac{TL}{GJ}$$

$$= \frac{45 \times 10^3}{160} \times \frac{3}{40 \times 10^9} \times \frac{32}{\pi \times 0.04677^4}$$

$$= 0.04487 \text{ rad}$$

$$\doteqdot 0.045 \text{ rad}.$$

5.2. A hollow shaft with a diameter ratio 3 : 5 is required to transmit 450 kW at 4π rad/s with a uniform torque. The shearing stress in the shaft must not exceed 60 MN/m², and the twist over a length of 3 m must not exceed 1°. Calculate the minimum external diameter of shaft satisfying these conditions. ($G = 80$ GN/m².) [London Univ.]

Solution

The selected shaft diameter must satisfy both conditions specified, i.e. the stress must not exceed 60 MN/m² and the twist must not exceed 1°. In such a problem the diameter must be calculated separately starting from each condition and the larger diameter selected.

(a) Diameter to meet maximum stress requirement.
From torsion formula,

$$\frac{T}{J} = \frac{q}{r}$$

and $P = T\omega$ relates power and torque.

$$\therefore T = \frac{P}{\omega}$$

$$= \frac{450\,000}{4\pi}$$

$$= 35\,810 \text{ Nm}.$$

For a hollow shaft $J = \dfrac{\pi}{32}[D^4 - d^4]$, where D is the outside and d the inside diameter.

Here $\quad d = \tfrac{3}{5}D$.

$$\therefore \quad J = 0.0855 D^4.$$

Maximum stress occurs at outside of shaft where

$$r = \frac{D}{2}.$$

$$\therefore \quad \frac{35\,801}{0.0855 D^4} = \frac{60 \times 10^6}{D/2},$$

$$D^3 = \frac{35\,801}{0.0855 \times 2 \times 60 \times 10^6},$$

$$D = 0.1515 \text{ m}.$$

(b) Diameter to meet twist condition,

$$\frac{T}{J} = \frac{G\theta}{L}.$$

Condition is that $\theta = 1/57.3$ rad for $L = 3$ m.

$$\therefore \quad D^4 = \frac{35\,810 \times 3 \times 57.3}{0.0855 \times 80 \times 10^9},$$

$$D = 0.1732 \text{ m}.$$

The smaller shaft would meet the stress but not the twist criterion; the larger shaft would meet both and is therefore the one required.

5.3. A composite shaft consists of a solid steel rod 75 mm diameter surrounded by a closely fitting brass tube firmly fixed to it. A torque is applied to the composite shaft half of which is carried by the steel rod and half by the brass tube. What must the outside diameter of the brass tube be? If the applied torque is 16 200 N m, calculate the maximum stress in each material and the angle of twist on a 4 m length. ($G_{\text{steel}} = 80$ GN/m². $G_{\text{brass}} = 40$ GN/m².)

[London Univ.]

Solution

Since the tube and rod are firmly fixed together, the angle of twist of each is the same. Using the suffix s for steel and b for brass, $\theta_s = \theta_b$.

But from the torsion formula, $\theta = TL/GJ$.

$$\therefore \frac{T_s L_s}{G_s J_s} = \frac{T_b L_b}{G_b J_b}.$$

Now $G_s = 2G_b$, $L_s = L_b$, $T_s = T_b$ in this problem.

$$\therefore J_b = 2J_s,$$

$$J_s = \frac{\pi D^4}{32} = \frac{\pi \times 0.075^4 \text{ m}^4}{32},$$

$$J_b = \frac{\pi}{32}(D^4 - 0.075^4)$$

but $\qquad J_b = 2J_s.$

$$\therefore \frac{\pi}{32}(D^4 - 0.075^4) = 2\frac{\pi}{32} \times 0.075^4,$$

$$D^4 = 3 \times 0.075^4,$$
$$D = 0.075 \sqrt[4]{3}$$
$$= 98.7 \text{ mm}.$$

$$T_s = T_b = \frac{16\,200}{2}$$
$$= 8100 \text{ N m}.$$

$$q_s = \frac{T_s r_s}{J_s}$$
$$= 8100 \times \frac{0.075}{2} \times \frac{32}{\pi \times 0.075^4}$$
$$= \frac{8100 \times 16}{\pi \times 0.075^3}$$
$$= \frac{8100 \times 16}{\pi \times 7.5^3} \times 10^6$$
$$= 97.7 \text{ MN/m}^2.$$

$$q_b = 8100 \times \frac{0.0987}{2} \times \frac{32}{\pi(0.0987^4 - 0.075^4)}$$
$$= 64.33 \text{ MN/m}^2.$$

Angle of twist $\theta = TL/GJ$, and is the same for both shafts. Calculating for steel gives

$$\theta = \frac{8100 \times 4 \times 32}{80 \times 10^9 \times \pi \times 0.075^4} \text{ rad}$$

$$= 7.43°.$$

5.4. A hollow steel shaft 100 mm external and 50 mm internal diameter is securely fixed at each end, the free length being 6 m. Twisting moments are applied to the shaft in the positions shown in Fig. 5.4. Find the end fixing moments, the maximum shearing stress and the greatest angle of twist of the shaft. ($G = 80$ GN/m².)

[London Univ.]

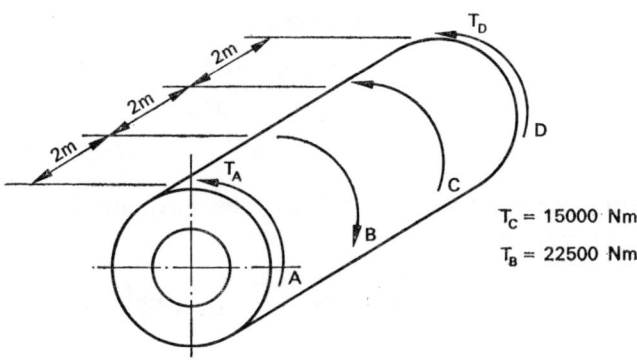

Fig. 5.4.

Solution

Assuming the fixing torques are T_A and T_D in the sense shown; considering the equilibrium of the shaft as a whole the sum of the torques must be zero,

i.e. $\qquad T_A - 22\,500 + 15\,000 + T_D = 0.$ \hfill (1)

Over each section of the shaft the twist relation $\theta = TL/GJ$ will apply and the twist of end D relative to end A will be $\Sigma TL/GJ$, but since both ends are fixed, the twist of D relative to $A = 0$.

$$\therefore \frac{T_A L_{AB}}{GJ} + \frac{(T_A - 22\,500) L_{BC}}{GJ} + \frac{(T_A + 15\,000 - 22\,500)}{GJ} L_{CD} = 0.$$

Now $L_{AB} = L_{BC} = L_{CD} = 2$ m.

$$\therefore T_A + T_A - 22\,500 + T_A + 15\,000 - 22\,500 = 0,$$
$$3 T_A = 30\,000,$$
$$T_A = 10\,000 \text{ N m}.$$

Substituting this in eqn. (1),

$$T_D = -2500 \text{ N m}.$$

The shearing stress $q = (Tr/J)$ and will be a maximum at the outer fibres in the section where T is greatest, i.e. between B and C.

$$q_{max} = \frac{12\,500 \times 0.05}{(\pi/32)(0.1^4 - 0.05^4)}$$
$$= 67.9 \text{ MN/m}^2.$$

Angle of twist $\theta = LT/GJ$, and this will be maximum in the section where T is maximum,

$$\theta = \frac{12\,500 \times 2 \times 32}{80 \times 10^9 \times \pi \times 0.9375 \times 10^{-4}} \text{ rad}$$
$$= \frac{1.25 \times 2 \times 32}{8 \times \pi \times 93.97} \times \frac{180}{\pi} \text{ degree}$$
$$= 1.945°.$$

5.5. A solid shaft of 50 mm diameter is connected to a hollow shaft of outside diameter 100 mm by means of a coupling with ten fitted bolts equi-spaced on a pitch circle of 250 mm diameter. The design of the coupling is such that shafts and coupling all have the same strength, and the allowable stresses are 50 MN/m² for the shaft and 25 MN/m² for the bolts.

Find the necessary diameter of the bolts and the internal diameter of the hollow shaft.

Solution

In order to be equally strong the shafts and bolts must all reach the maximum allowable stress at the same torque. It is easy to see that if any one of the three reached its maximum allowable stress before the others, then the others would be unnecessarily strong and a more economical design could be achieved which would carry the same torque.

Torque carried by solid shaft $T_s = q\dfrac{J}{r}$,

$$J = \frac{\pi D_s^4}{32}, \quad r = \frac{D_s}{2}, \quad T_s = q\frac{\pi D_s^3}{16},$$

but q is limited to 50 MN/m².

$$\therefore\ T_s = \frac{50\pi \times 10^6}{16} D_s^3,$$

where $D_s = 0.05$ m.

Torque carried by hollow shaft $T_H = q\dfrac{\pi}{32}(D^4 - d^4)\dfrac{2}{D}$, q is limited to 50 MN/m² and $D = 0.1$ m.

$$T_H = \frac{50\pi}{16}\frac{(0\cdot1^4 - d^4)}{0\cdot1}\times 10^6.$$

The same torque is carried by each shaft.

$$\therefore\ \frac{50\pi \times 10^6}{16}\times 0\cdot05^3 = \frac{50\pi}{16}\frac{(0\cdot1^4 - d^4)}{0\cdot1}\times 10^6,$$

$$d^4 = 0\cdot1^4 - 0\cdot05^3 \times 0\cdot1$$

$$= 10^{-4}(1 - 0\cdot125)$$

$$d = 10^{-1}\sqrt[4]{(0\cdot875)}$$

$$= 0\cdot967 \text{ m}$$

$$= 96\cdot7 \text{ mm}.$$

122   STRESS ANALYSIS PROBLEMS IN S.I. UNITS

Torque carried by bolts = shear stress in bolts × area of bolt × number of bolts × radius of pitch circle,

$$T_B = qA \frac{d_b}{2} N.$$

This is the same as the torque carried by the shafts.

$$\therefore \frac{50\pi \times 10^6}{16} \times 0.05^3 = 25 \times 10^6 \times \frac{\pi d_b^2}{4} \times 10 \times 0.125,$$

where d_b is the diameter of a bolt.

$$d_b^2 = \frac{50 \times 125}{16} \times \frac{4 \times 10^{-6}}{25 \times 1.25}$$

$$= \frac{200}{4} \times 10^{-6},$$

$$d_b = 10^{-3} \sqrt{50} \text{ m},$$

$$d_b = 7.1 \text{ mm}.$$

5.6. A hollow shaft 100 mm external diameter and 50 mm internal diameter transmits 0·6 MW at 50 rad/s and is subjected to an end thrust of 45 kN. Find what bending moment may be safely applied to the shaft if the greater principal stress is not to exceed 90 MN/m².

[London Univ.]

Solution

Power in watts, $P = T\omega$,

$$T = \frac{P}{\omega},$$

$$= \frac{600\,000}{50} = 12\,000 \text{ N m}.$$

$$J = \frac{\pi}{32}(D^4 - d^4)$$

$$= 9.206 \times 10^{-6} \text{ m}^4.$$

At shaft surface shear stress due to torsion,

$$q = \frac{0.1 \times 12\,000 \times 10^6}{2 \times 9.206}$$

$$= 65.18 \text{ MN/m}^2.$$

The greater principal stress is 90×10^6 N/m².

The stress system at the upper surface of the shaft will be that shown in Fig. 5.3b in definitions and theory section (f), with the compressive stress marked B being due to the end thrust and the bending effect.

The general equation for the principal stresses in a complex stress system is

$$f_{1,2} = \tfrac{1}{2}(f_x + f_y) \pm \tfrac{1}{2}\sqrt{[(f_x - f_y)^2 + 4q_{xy}^2]}.$$

In the system illustrated, $f_x = -f_B$, $f_y = 0$, $q_{xy} = q$.

$$\therefore f_{1,2} = -\tfrac{1}{2}f_B \pm \tfrac{1}{2}\sqrt{(f_B^2 + 4q^2)}.$$

The greater principal stress will be f_2, the compressive stress

$$-90 \times 10^6 = -\tfrac{1}{2}f_B - \tfrac{1}{2}\sqrt{[f_B^2 + 4 \times (65.18 \times 10^6)^2]}$$
$$-4(-90 \times 10^6 + \tfrac{1}{2}f_B)^2 = f_B^2 + 4(65.18 \times 10^6)^2.$$

Solving the resulting quadratic gives $f_B = 42.78$ MN/m². Of this the stress due to direct load is

$$\frac{45\,000}{(\pi/4)(D^2 - d^2)} = 7.638 \text{ MN/m}^2.$$

$$\therefore \text{ bending stress} = 42.78 - 7.638 \text{ MN/m}^2$$
$$= 35.14 \times 10^6 \text{ N/m}^2.$$

$$M = \frac{fI}{y} = \frac{35.14 \times 10^6}{0.05} \times \frac{\pi}{64}(D^2 + d^2)(D^2 - d^2)$$

$$= 3235 \text{ N m}.$$

Problems

5.7. A solid shaft is required to transmit 70 kW at 11 rad/s with uniform torque.

(a) What is the minimum diameter shaft needed if shearing stress must not exceed 70 MN/m²?

(b) What is the minimum diameter of shaft needed if the twist in a length of 2 m is not to exceed 2°?

(c) If both the above conditions apply at the same time, what diameter must be used?

[(a) 79 mm. (b) 82·6 mm. (c) 82·6 mm.]

5.8. A 32 mm diameter solid shaft forms the transmission in the drill of an oil well 3350 m deep. The drilling speed is 50 rad/s and the maximum shear stress is limited to 48 MN/m². Find (a) the limiting value of the torque which can be transmitted, (b) the angle of twist of one end of the shaft relative to the other (the modulus of rigidity is 80 GN/m²), and (c) the power required for the drive.

[London Univ.]

[(a) 308·8 N m. (b) 126 rad. (c) 15·44 kW.]

5.9. A vertical drilling rod for an oil-well rig is driven by a torque at the top end against a resistance concentrated at the bottom end. If the maximum shearing stress in the rod has a limiting value of 80×10^6 N/m² when drilling at a depth of 3000 m, find the angular rotation of one end of the rod relative to the other. The diameter of the shaft is 100 mm and $G = 80$ GN/m².

If drilling is carried out under the same conditions with a hollow shaft of the same material 125 mm outside diameter and 75 mm inside diameter, what would be the maximum shearing stress and the total angle of twist? [London Univ., 1967]

[(a) 60 rad. (b) 47·1 MN/m². (c) 28·24 rad.]

TORSION

5.10. A uniform circular steel rod is held at end A. At the other end C a torque of 1000 N m is applied and at the mid-point B of the length a torque of 250 N m is applied in the sense opposite to that at C. Specify the torque acting at various points in the rod. If the shear stress is not to exceed 75 MN/m², what must the diameter be?

[London Univ., 1966]

[A to B 750 N m. B to C 1000 N m. Diameter = 40·8 mm.]

5.11. A vessel having a single propeller shaft of 300 mm diameter and running at $16\pi/3$ rad/s is re-engined with turbines driving two equal propeller shafts at 50π rad/s and developing 60 per cent more power. If the working stresses in the new shafts are 10 per cent greater than in the old shaft, find the diameters of the new shafts.

[London Univ.]

[161 mm.]

5.12. A solid circular shaft of 250 mm diameter is to be replaced by a hollow shaft of the same material, the ratio of the external to internal diameters being 2 : 1. Find the size of the hollow shaft if the maximum shear stress is the same for both shafts. Calculate also the ratio of weights of the hollow to solid shaft.

[Hollow shaft o.d. = 255·5 mm; i.d. = 127·75 mm. Weight of hollow : weight of solid = 0·782 : 1.]

5.13. A steel rod is surrounded by a closely fitting duralumin tube and the two are securely fastened together to form a composite shaft. Find the diameter of the steel rod and the outside diameter of the duralumin tube so that the maximum shearing stresses in the two materials do not exceed 92·5 MN/m² and 60 MN/m² respectively when the composite shaft is subjected to a torque of 700 N m. Also calculate the angle of twist in a length of 1·25 m. (G for duralumin = 26 GN/m²; G for steel = 80 GN/m².) [London Univ.]

[18·75 mm. 37·5 mm. 8·83°.]

5.14. A steel bar 20 mm diameter is encased in a closely fitting brass tube 32 mm external diameter; both are securely fixed together at the ends to form a compound bar. A torque of 520 N m is applied and the angle of twist measured on a gauge length of 25 mm is found to be 1·8°. If G for the steel is assumed to be 80 GN/m², calculate G for the brass. Find also the maximum shearing stresses in the two materials. [London Univ.]

[32·8 GN/m². Maximum stress in steel = 100 MN/m²; in brass 66·4 MN/m².]

5.15. A pulley can be fastened to a shaft either by a square key 6 mm by 6 mm by 24 mm long, or by a parallel pin 10 mm diameter (Fig. 5.5). If both key and pin can have the same maximum shear stress of 70 MN/m², which of these two fastenings will carry the greater torque between pulley and shaft? What torque will each carry?

[Pin carries 137·5 N m. Key carries 126 N m.]

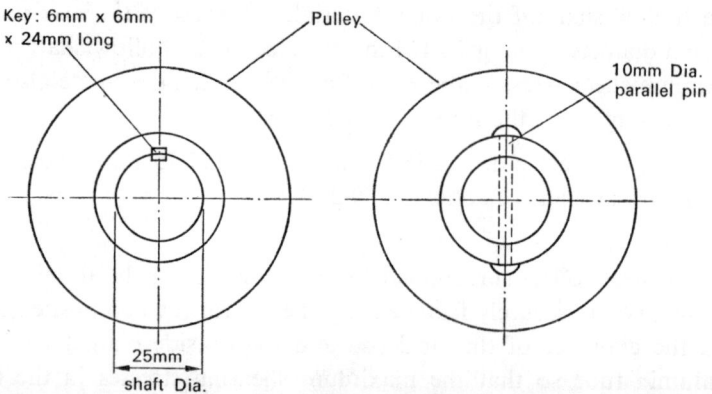

Fig. 5.5.

5.16. Two 50 mm solid shafts are coupled together by a flanged coupling with ten bolts on a pitch circle of 125 mm diameter. If the maximum shear stress in the shaft is to be 50 MN/m² and in the bolts

TORSION

is to be 25 MN/m², what must be the bolt diameter for both bolts and shaft to be equally strong?

[10 mm.]

5.17. A horizontal shaft AB securely fixed at each end has a free length of 9 m. Looking from end A, axial couples of 30 kN m clockwise at 3·5 m from A and 38 kN m counter-clockwise at 6 m from A, act on the shaft. Find the end fixing couples in magnitude and direction and find the diameter that a solid shaft must have if the maximum shearing stress is limited to 60 MN/m².

Draw a diagram to show how a line on the outer surface of the shaft, originally parallel to the axis, will appear. Find the position where the shaft suffers zero twist. [London Univ.]

[T_A = 5666 N m counter-clockwise viewed from end A. T_B = 13 666 N m clockwise. D = 127·5 mm. Point of zero twist is at 4·315 m from A.]

5.18. A solid shaft decreases in diameter from D to d uniformly over a length L. If it is under the action of a pure torque T, show that the total angle of twist is given by

$$\theta = \frac{32T}{3\pi G}\left[\frac{L}{D-d}\right]\left[\frac{1}{d^3}-\frac{1}{D^3}\right].$$

Find the angle of twist for a steel shaft 2 m long which tapers uniformly from 50 mm diameter to 25 mm diameter and carries a pure torque of 140 N m. Find also the maximum shear stress. (G = 80 GN/m².)

[1·525°. 45·6 MN/m².]

5.19. A shaft 1·5 m long tapers uniformly from 100 mm diameter at one end to 75 mm diameter at the other. The larger end is firmly fixed and a torque of 3400 N m is applied to the smaller end. Find the total angle of twist and the maximum shear stress. (G = 80 GN/m².)

[London Univ.]

[0·68°. 41 MN/m² (at the small end).]

5.20. A hollow marine propeller shaft turns at 11·5 rad/s. It is required to propel the vessel at 25 knots (12·875 m/s) by delivering 6·3 MW to a propeller which is 68 per cent efficient. The diameter ratio of the shaft is 2 : 3 and the direct stress from end thrust is not to exceed 7·75 MN/m².

Calculate the shaft diameters and the maximum shearing stress arising from the torque.

[310 mm and 207 mm. 116 MN/m².]

5.21. In a circular shaft subjected to an axial twisting moment T and a bending moment M, show that when $M = 1·2T$ the ratio of the maximum shearing stress to the greatest principal stress is approximately 0·566. [London Univ.]

5.22. A mild-steel shaft 150 mm diameter is subjected simultaneously to a torque of 14 kN m and a bending moment of 11 kN m. Find the maximum principal stress and also the maximum shear stress.

[43·45 MN/m². 26·85 MN/m².]

5.23. A solid shaft 50 mm diameter is securely fixed at either end of its 6 m length. A torque of 1500 N m is applied at a section 2 m from one end. ($G = 80$ GN/m².)

Calculate (a) the fixing torques, (b) the maximum shear stress in each section, and (c) the angle of twist at the section where the torque is applied.

[(a) 1000 N m and 500 N m. (b) 40·75 MN/m² and 20·375 MN/m². (c) 2·33°.]

5.24. Find the dimensions of a hollow shaft, i.d./o.d. = 0·6, which is to transmit 150 kW at 26 rad/s if the shearing stress is not to exceed 70 MN/m².

If a bending moment of 2700 N m is now applied to the shaft, find the speed at which it must be driven to transmit the same power for the same value of maximum shearing stress. [London Univ.]

[o.d. = 78·43 mm. i.d. = 47·06 mm. ω = 28·54 rad/s.]

TORSION

5.25. A flywheel weighing 5·4 kN is mounted on a shaft 75 mm diameter midway between two bearings 600 mm apart. If the shaft is transmitting 30 kW at 12π rad/s, calculate the principal stresses and the maximum shear stress at the ends of a vertical and horizontal diameter in a plane close to the flywheel. [London Univ.]

[At top of vertical diameter: principal stresses $+3\cdot922$ MN/m², $-23\cdot47$ MN/m²; maximum shear stress 13·7 MN/m². At bottom vertical diameter: principal stresses $+23\cdot47$ MN/m², $-3\cdot922$ MN/m²; maximum shear stress 13·7 MN/m². At either end of horizontal diameter: principal stresses $\pm9\cdot6$ MN/m²; maximum shear stress 9·6 MN/m².]

5.26. A hollow shaft of 0·2 m outside diameter and 0·125 m bore is subjected simultaneously to a bending moment of 43 kN m and a torque of 65 kN m. Calculate the maximum bending stress and the maximum torsional shear stress. Hence find the maximum shearing stress in the shaft. [London Univ.]

[58·53 MN/m².]

5.27. Show that, for a solid circular shaft of diameter d carrying a torque T and a bending moment M, the maximum shear stress in the cross-section is

$$q_M = \frac{16}{\pi d^3} \sqrt{(M^2+T^2)}$$

while the principal stresses are

$$f_{1,\,2} = \frac{16}{\pi d^3}[M \pm \sqrt{(M^2+T^2)}].$$

5.28. A steel shaft $ABCD$ of circular section is 1·9 m long and is supported in bearings at A and D. $AB = 0\cdot7$ m, $BC = 0\cdot5$ m, $CD = 0\cdot7$ m. The shaft is horizontal and two horizontal arms rigidly connected to it at B and C project at right angles on opposite sides. Arm B carries a vertical load of 18 kN at 0·3 m from the shaft axis and C

carries a vertical balancing load at 0·4 m from the shaft axis. If q_M is not to exceed 75×10^6 N/m², find the minimum diameter of shaft assuming the bearings give simple support. [London Univ.]

[95 mm.]

5.29. A hollow steel shaft having an internal diameter to external diameter ratio of 0·6 is required to transmit 5400 kW when rotating at a constant speed of 300 rev/min. Given that the shear stress is not to exceed 55 MN/m², find the minimum external diameter of the shaft. What is the minimum shear stress in the shaft? The shear modulus for the material of the shaft is 79·3 GN/m². Determine the angle of twist (in degrees) of a length of the shaft equal to twenty times the external diameter when transmitting the required power.

[London Univ., 1969]

[263·5 mm. 33 MN/m². 1·59°.]

CHAPTER 6

Strain Analysis

Definitions and Theory

(a) In practice one cannot measure stress in a material directly but one can measure strain (e.g. by using strain gauges) and deduce the stress.

(b) Direct strain in a particular direction (symbol e with a suffix to indicate direction) is defined as the ratio of change in length to original length in that direction. Shear strain (symbol γ with two suffixes to indicate the planes concerned) is defined as the change in angle (in radians) between two planes originally at right angles.

(c) Plane strain is the term used to describe a strain system when the direct strain in one direction, and the two corresponding shear strains, is zero (i.e. $e_z = 0$, $\gamma_{zx} = 0$, $\gamma_{zy} = 0$).

(d) An element of material such as that shown in Fig. 6.1 which has direct strains e_x and e_y in the x and y directions respectively and a shear strain γ_{xy} will have direct and shear strains associated with the plane at θ counter-clockwise to face OP of

$$e_n = \tfrac{1}{2}(e_x+e_y)+\tfrac{1}{2}(e_x-e_y)\cos 2\theta + \tfrac{1}{2}\gamma_{xy}\sin 2\theta,$$
$$\tfrac{1}{2}\gamma_s = \tfrac{1}{2}(e_x-e_y)\sin 2\theta - \tfrac{1}{2}\gamma_{xy}\cos 2\theta.$$

Note. The above expressions are similar to those for stresses on oblique planes except for the fact that the coefficient $\tfrac{1}{2}$ is associated with all the shear-strain terms while no such coefficient is associated with shear-stress terms in the stress equations.

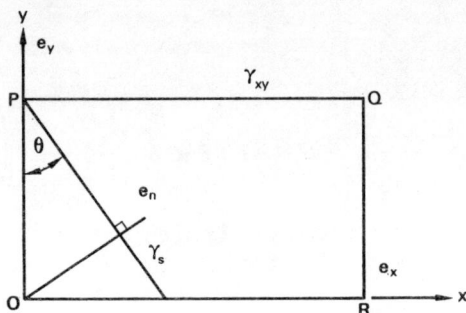

Fig. 6.1. Strains on a plane.

(e) Mohr's strain circle is a graphical method of solving problems concerning strain at a point. The construction is like that of the stress circle (Chapter 3) except that semi-shear strains ($\frac{1}{2}\gamma$) are plotted vertically. The sign conventions are chosen to correspond to those for stresses; thus tensile strains are positive and compressive strains are negative. For shear strains clockwise rotation of a plane implies a negative strain and counter-clockwise rotation a positive strain. (N.B.—This convention may seem to be opposite to that for stresses but it does, in fact, ensure that a plane carrying a positive shear stress undergoes a positive shear strain.)

(f) In a three-dimensional stress system with stresses f_x, f_y, and f_z, q_{xy}, q_{yz}, and q_{zx} and their complementaries all acting, the relations between stresses and strains are obtained from the equations

$$e_x = \frac{1}{E}[f_x - \nu(f_y + f_z)] \qquad \gamma_{xy} = \frac{q_{xy}}{G},$$

$$e_y = \frac{1}{E}[f_y - \nu(f_z + f_x)] \qquad \gamma_{yz} = \frac{q_{yz}}{G},$$

$$e_z = \frac{1}{E}[f_z - \nu(f_x + f_y)] \qquad \gamma_{zx} = \frac{q_{zx}}{G}.$$

(g) It can be shown that the elastic moduli E, G, and K and Poisson's

STRAIN ANALYSIS

ratio v are connected by the three equations

$$E = 3K(1-2v),$$
$$E = 2G(1+v),$$
$$3K(1-2v) = 2G(1+v).$$

Worked Examples

6.1. On an element of material the total direct strain in the x direction is e_x in magnitude and it is tensile. In the y direction the strain is also tensile and of magnitude e_y ($e_y < e_x$). The magnitude of the shear strain is γ_{xy} on the y face and it is positive. Sketch Mohr's circle for the system and from the circle derive expressions for the magnitudes of the maximum and minimum direct strains and the maximum shear strain.

Solution

Referring to Fig. 6.2.

In plotting the circle it is essential to note that semi-shear strains are plotted, i.e. $\frac{1}{2}\gamma_{xy}$, and that if this is positive with respect to the y

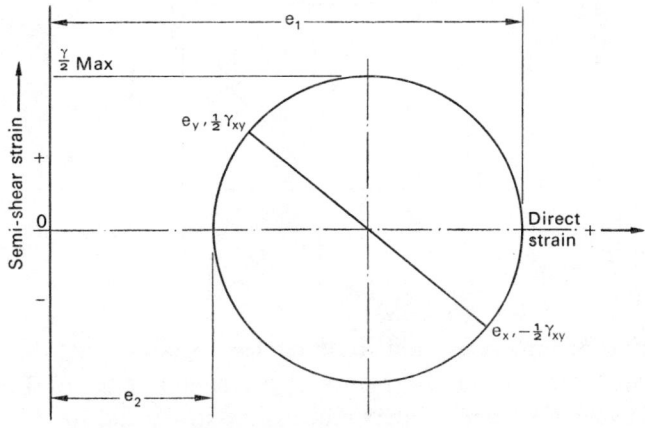

FIG. 6.2.

face as the question states, then it will be negative with respect to the x face but of the same magnitude.

Coordinates of centre of circle $\frac{1}{2}(e_x+e_y)$, 0.

$$\text{Radius of circle} = \sqrt{\left[\left(\frac{e_x-e_y}{2}\right)^2+\left(\frac{\gamma_{xy}}{2}\right)^2\right]}$$

$$= \frac{1}{2}\sqrt{[(e_x-e_y)^2+(\gamma_{xy})^2]}.$$

Maximum shear strain = $2\times$radius = γ_{max}

$$\gamma_{max} = \sqrt{[(e_x-e_y)^2+(\gamma_{xy})^2]}.$$

Maximum direct strain = $e_1 = \frac{1}{2}(e_x+e_y)+\frac{1}{2}\sqrt{[(e_x-e_y)^2+(\gamma_{xy})^2]}$.

Minimum direct strain = $e_2 = \frac{1}{2}(e_x+e_y)-\frac{1}{2}\sqrt{[(e_x-e_y)^2+(\gamma_{xy})^2]}$.

6.2. At a point in a member the stress system is as shown in Fig. 6.3. The stress f_x is to be assumed greater than f_y. In the x direction the tensile strain is of magnitude e_x and in the y direction there is also a tensile strain of magnitude e_y ($e_y < e_x$). The shear strain related to the y face is γ_{xy} (positive).

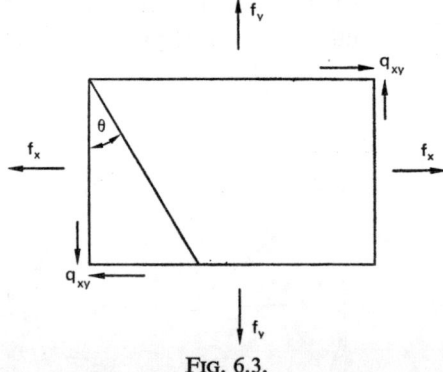

Fig. 6.3.

(a) Draw Mohr's stress and strain circles. (b) Show that the planes of principal stresses are also planes of principal strains. (c) Find the angle between the plane of maximum shear strain and the first principal plane.

Solution

(a) The stress and strain circles are shown in Fig. 6.4.

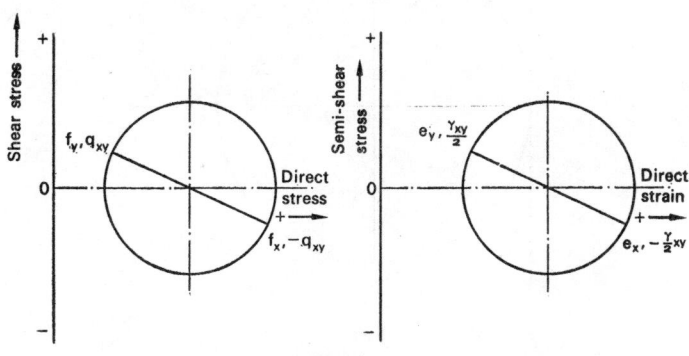

Fig. 6.4.

(b) The relation between shear stress and shear strain relative to any plane is $q/\gamma = G$. Thus if $q = 0$, $\gamma = 0$. From inspection of the circles above it is clear that there are only two points where shear stresses are zero and two where shear strains are zero. It is also clear that shear stress is zero on a plane of principal stress and shear strain is zero on a plane of principal strain. Therefore the planes of principal stresses are the same as the planes of principal strains.

(c) 45°.

6.3. Three strain gauges are fastened in the same plane to a body which is stressed only in the plane of the gauges. Gauge 2 is at 45° and gauge 3 at 90° to gauge 1. If gauge 1 reads +0·000125, gauge 2 reads +0·00025 and gauge 3 reads +0·000125, what are the principal stresses in the body? ($E = 200$ GN/m². $\nu = 0·3$.)

Solution

Refer to Fig. 6.5. Since gauges 1 and 3 are 90° apart on the body they will plot 180° apart on the strain circle. Since both read the same value they must lie at opposite ends of the vertical diameter and therefore

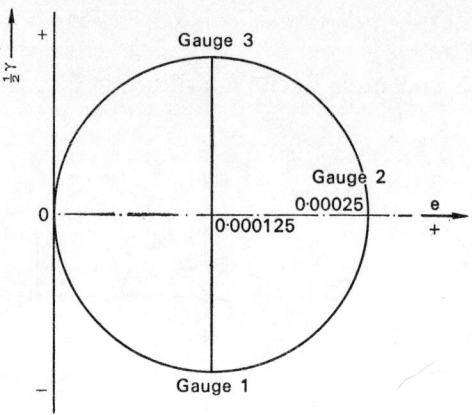

Fig. 6.5.

the circle centre is at $+0 \cdot 000125$. As the intermediate gauge reads $+0 \cdot 00025$, this must be the value of the greatest principal strain e_1.

A principal strain is caused by the stress in its own direction and by the Poisson effects due to the other two principal stresses. In the present problem $f_3 = 0$.

$$\therefore e_1 = \frac{1}{E}(f_1 - \nu f_2),$$

$$e_2 = \frac{1}{E}(f_2 - \nu f_1).$$

From the circle, Fig. 6.5,

$$e_2 = 0.$$

$$\therefore f_2 = \nu f_1 \quad \text{and} \quad e_1 E = f_1 - \nu^2 f_1.$$

But $e_1 = 25 \times 10^{-5}$; $E = 2 \times 10^{11}$, and $\nu = 3 \times 10^{-1}$.

$$\therefore 25 \times 10^{-5} \times 2 \times 10^{11} = f_1(1 - 0 \cdot 09),$$

$$f_1 = \frac{5 \times 10^7}{0 \cdot 91} \text{ N/m}^2$$

$$= 55 \text{ MN/m}^2.$$

$$f_2 = 0 \cdot 3 \times 55 = 16 \cdot 5 \text{ MN/m}^2.$$

6.4. Three strain gauges are arranged at 60° to one another to form a rosette. The gauges read strains of 0·00046, 0·0002, and −0·00016. (a) Using Mohr's strain circle, and (b) without using Mohr's strain circle, find the values of maximum and minimum principal stresses and strains. ($E = 207$ GN/m². $v = 0.29$.)

Solution

Method (a)

Draw a vertical line for the semi-shear strain axis and to a suitable scale set off the vertical lines *AB*, *CD*, and *EF* shown in Fig. 6.6 to locate the direct strain coordinates read by the gauges. At this stage the horizontal axis cannot be located since the semi-shear strains corresponding to the direct strains are unknown.

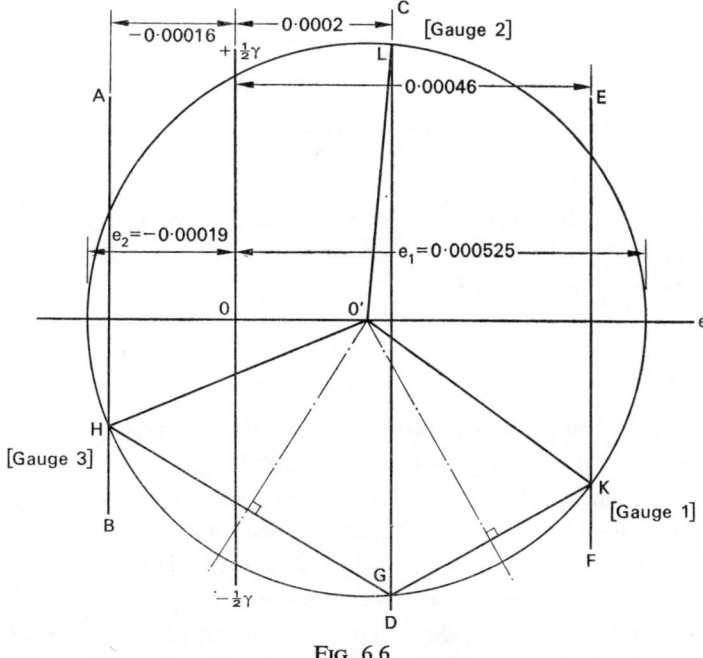

Fig. 6.6.

At any point G on the middle of the three direct strain lines—CD in this case—draw lines GH and GK to make the angles CGH and CGK equal to the rosette angle of 60° and to intersect AB and EF at H and K respectively.

Draw in the perpendicular bisectors of GH and GK and from their intersection O' draw a circle to pass through points H, G, and K. The horizontal through O' is then the direct strain axis, and the direct and semi-shear strains associated with planes perpendicular to the lines of the gauges are then the co-ordinates of the points H, K, and L, where L is at the point where the circle crosses vertical CD.

From the circle:

$e_1 = 0.000525$ at 17.5° counter-clockwise from gauge 1,

$e_2 = -0.00019$ at 107.5° counter-clockwise from gauge 1.

Now
$$e_1 = \frac{1}{E}(f_1 - \nu f_2),$$

$$e_2 = \frac{1}{E}(f_2 - \nu f_1).$$

Substituting values for e_1, e_2, E, and ν and solving simultaneously gives $f_1 = 106$ MN/m²,
$f_2 = -8.56$ MN/m².

Method (b)

The gauge reading 0.00046 must be nearest to the line of the maximum principal strain since it has the greatest reading. Assuming that the angle between its direction and that of e_1 is α, then the gauge reading -0.00016 is at $\alpha + 60°$, and the gauge reading 0.0002 is at $\alpha + 120°$. (Reference to Fig. 6.6 will verify these angles if it is remembered that angles are doubled when one moves to the circle plot.)

The equations relating direct strains to principal strains are given in item (d) of the definitions and theory section, and substituting in the first of these gives three equations:

$$0\cdot 00046 = \tfrac{1}{2}(e_1+e_2)+\tfrac{1}{2}(e_1-e_2)\cos 2\alpha, \quad (1)$$
$$-0\cdot 00016 = \tfrac{1}{2}(e_1+e_2)+\tfrac{1}{2}(e_1-e_2)\cos(2\alpha+120°), \quad (2)$$
$$0\cdot 0002 = \tfrac{1}{2}(e_1+e_2)+\tfrac{1}{2}(e_1-e_2)\cos(2\alpha+240°). \quad (3)$$

Note that the term $\tfrac{1}{2}\gamma_{xy}\sin 2\theta$ has disappeared because the equations are written now relative to the principal stress directions for which γ is zero.

Subtracting eqn. (2) from eqn. (1) and using the identity,

$$\cos A - \cos B \equiv 2\sin\tfrac{1}{2}(A+B)\sin\tfrac{1}{2}(B-A)$$

gives

$$0\cdot 00062 = \tfrac{1}{2}(e_1-e_2)\, 2\sin\tfrac{1}{2}(4\alpha+120°)\sin 60°,$$
$$0\cdot 00062 = \frac{\sqrt{3}}{2}(e_1-e_2)\sin(2\alpha+60°). \quad (4)$$

Subtracting eqn. (3) from eqn. (2) and using the same substitution of identity gives

$$-0\cdot 00036 = \tfrac{1}{2}(e_1-e_2)\, 2\sin(2\alpha+180°)\sin 60°,$$
$$-0\cdot 00036 = \tfrac{1}{2}\sqrt{(3)}\,(e_1-e_2)\sin(2\alpha+180°),$$
$$-0\cdot 00036 = -\tfrac{1}{2}\sqrt{(3)}\,(e_1-e_2)\sin 2\alpha, \quad (5)$$

since $\sin(2\alpha+180°) = -\sin 2\alpha$.

Dividing eqn. (4) by eqn. (5) and using the identity

$$\sin(A+B) \equiv \sin A\cos B + \cos A\sin B,$$
$$1\cdot 725 = \sin 60°\cot 2\alpha + \cos 60°,$$
$$\cot 2\alpha = 1\cdot 413.$$
$$\therefore \alpha = 17°\,39'.$$

This value can now be substituted back in any two of equations (1) to (3) and the pair solved simultaneously to give

$$e_1 = 0\cdot 000525,$$
$$e_2 = -0\cdot 000192.$$

The angles to gauge 1 are

$$17° \; 39' \quad \text{and} \quad 107° \; 39'.$$

The stresses can now be calculated as in the solution to (a).

(*Note:* The very great saving of labour in solution (a) compared with solution (b) should impress upon the student the advantage of mastering Mohr's circle methods for both strain and stress problems.)

6.5. Derive the relationships between the elastic constants (a) E, K, and ν, and (b) E, G, and ν, and hence express K in terms of E and G.
[London Univ., B.Sc. 2]

Solution

(a) Consider a rectangular block under uniform hydrostatic pressure p. Total strain along any side of the rectangle,

$$e = \frac{1}{E}(p - \nu(p+p))$$

$$= \frac{p}{E}(1 - 2\nu).$$

Volumetric strain $= \dfrac{3p}{E}(1 - 2\nu).$

But volumetric strain $= \Delta V / V$

and by definition $K = \dfrac{P}{\Delta V / V}.$

$$\therefore \quad \frac{p}{\dfrac{3p}{E}(1-2\nu)} = K.$$

$$\therefore \quad E = 3K(1 - 2\nu).$$

(b) In a rectangular element of material carrying the general stress system shown in Fig. 6.7, a plane at θ to the x face is chosen

STRAIN ANALYSIS

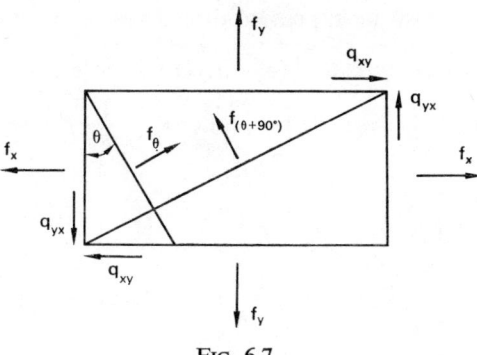

FIG. 6.7.

with θ such that the plane is perpendicular to the diagonal of the element. The direct stress on this plane will be f_θ and on the diagonal plane $f_{\theta+90°}$.

$$f_\theta = \tfrac{1}{2}(f_x+f_y)+\tfrac{1}{2}(f_x-f_y)\cos 2\theta + q_{xy}\sin 2\theta,$$

$$f_{(\theta+90°)} = \tfrac{1}{2}(f_x+f_y)+\tfrac{1}{2}(f_x-f_y)\cos 2(90°+\theta)+q_{xy}\sin 2(90°+\theta),$$

$$f_{(\theta+90°)} = \tfrac{1}{2}(f_x+f_y)-\tfrac{1}{2}(f_x-f_y)\cos 2\theta - q_{xy}\sin 2\theta.$$

Strain in direction of $f_\theta = e_\theta$, where

$$e_\theta = \frac{1}{E}(f_\theta - \nu f_{(\theta+90°)}).$$

$$e_\theta = \frac{1}{E}\left\{\frac{1}{2}(f_x+f_y)+\frac{1}{2}(f_x-f_y)\cos 2\theta + q_{xy}\sin 2\theta \right. $$
$$\left. -\nu\left[\frac{1}{2}(f_x+f_y)-\frac{1}{2}(f_x-f_y)\cos 2\theta - q_{xy}\sin 2\theta\right]\right\},$$

$$e_\theta = \frac{1}{E}\left[\frac{1}{2}(f_x+f_y)(1-\nu)+\frac{1}{2}(f_x-f_y)(1-\nu)\cos 2\theta \right.$$
$$\left. + q_{xy}(1+\nu)\sin 2\theta\right]. \tag{1}$$

Now e_θ can also be expressed directly in terms of e_x, e_y, and γ_{xy}:

$$e_\theta = \tfrac{1}{2}(e_x+e_y)+\tfrac{1}{2}(e_x-e_y)\cos 2\theta+\tfrac{1}{2}\gamma_{xy}\sin 2\theta,$$

$$e_x = \frac{1}{E}(f_x-\nu f_y) \quad \text{and} \quad e_y = \frac{1}{E}(f_y-\nu f_x).$$

$$\therefore\ e_\theta = \frac{1}{2}\left[\frac{1}{E}(f_x-\nu f_y)+\frac{1}{E}(f_y-\nu f_x)\right]$$

$$+\frac{1}{2}\left[\frac{1}{E}(f_x-\nu f_y)-\frac{1}{E}(f_y-\nu f_x)\right]\cos 2\theta+\frac{1}{2}\gamma_{xy}\sin 2\theta$$

$$e_\theta = \frac{1}{E}\left[\frac{1}{2}(f_x+f_y)(1-\nu)+\frac{1}{2}(f_x-f_y)(1-\nu)\cos 2\theta\right]$$

$$+\frac{1}{2}\gamma_{xy}\sin 2\theta. \tag{2}$$

Subtracting eqn. (2) from eqn. (1),

$$0 = \frac{1}{E}q_{xy}(1+\nu)\sin 2\theta - \frac{1}{2}\gamma_{xy}\sin 2\theta,$$

$$\frac{1}{2}\gamma_{xy} = \frac{1}{E}q_{xy}(1+\nu)$$

$$\frac{E}{2(1+\nu)} = \frac{q_{xy}}{\gamma_{xy}}. \quad \text{But} \quad \frac{q_{xy}}{\gamma_{xy}} = G.$$

$$\therefore\ E = 2G(1+\nu),$$

$$\nu = \frac{E}{2G}-1,$$

and substituting in the relation from (a),

$$E = 3K\left(1-\frac{E}{G}+2\right),$$

$$\frac{EG}{3(3G-E)} = K.$$

STRAIN ANALYSIS

6.6. A cylindrical boiler, 1·5 m diameter, has two strain gauges on the outer surface, one of which measures the hoop strain and the other the longitudinal strain. The boiler has a plate thickness of 10 mm and may be assumed to act like a theoretical thin cylinder with closed ends. When pressure is applied the hoop strain gauge indicates a strain of 0·0005. If $E = 200$ GN/m² and $v = 0·3$, find (a) the pressure in the boiler, and (b) the reading of the other gauge.

Solution

If e_H and f_H are the hoop strain and stress and f_L and e_L the longitudinal stress and strain,

$$e_H = \frac{1}{E}(f_H - vf_L) \quad \text{and} \quad e_L = \frac{1}{E}(f_L - vf_H).$$

In a thin cylinder, $f_H = 2f_L$.

$$\therefore e_H = \frac{f_H}{E}\left(1 - \frac{v}{2}\right).$$

Substituting values gives $f_H = \dfrac{200}{1·7}$ MN/m².

But
$$f_H = \frac{pr}{t}.$$

$$\therefore p = 1·57 \text{ MN/m}^2.$$

$$e_L = \frac{1}{E}(f_L - vf_H),$$

$f_H = 175$ MN/m², $f_L = 87·5$ MN/m².

$$\therefore e_L = 0·000175.$$

Problems

6.7. The strains across the x and y planes in a steel member were found to be $e_x = 0·00050$, $e_y = 0·00014$, $\gamma_{yx} = +0·00036$. Construct Mohr's strain circle and find the magnitudes and directions of the principal strains.

(*Note:* Just as q_{yx} is the shear stress on the same face as direct stress f_x, so γ_{yx} is the shear strain associated with direct strain e_x, while γ_{xy} is associated with e_y.)

[$e_1 = 0.0005745$ across a plane at $-22° 30'$ to the x plane. $e_2 = 0.0000655$ across a plane at $-112° 30'$ to the x plane.]

6.8. Draw Mohr's circle and find the magnitudes and directions of the principal strains for the point in a member where the strains are $e_x = 0.00050$, $e_y = +0.00030$, and $\gamma_{xy} = +0.00105$.

[$e_1 = 0.000934$. $\theta_1 = 39° 36'$. $e_2 = -0.000134$. $\theta_2 = 129° 36'$.]

6.9. At a point on the surface of a shaft there is an axial strain due to a compressive end load of -0.0003 and a shear strain on the cross-sectional plane of $+0.0004$. If $\nu = 0.3$, find the values of e_1, e_2, and γ_{max}.

[$e_1 = 0.000175$. $e_2 = -0.000385$. $\gamma_M = 0.00056$.]

6.10. Two strain gauges are fitted to one side of a high-tensile steel bar of rectangular cross-section, area 400 mm². The gauges are at 45°, one on either side of the centre line. What strains will each indicate if a tensile axial load of 200 kN is applied? ($E = 200$ GN/m². $\nu = 0.3$.)

[$+0.000875$.]

6.11. A shaft of 50 mm diameter carries bending and torsion. The principal strains measured at the shaft surface are $e_1 = +0.0011$, $e_2 = -0.0006$, with e_1 inclined at 20° to the shaft axis. If G for the material is 84 GN/m², find the maximum shear stress in the material and the applied torque.

[142·9 MN/m². 2250 N m.]

6.12. Three strain gauges are arranged in the form of a rosette with the angle between first and second of 45° and between first and third of 90°.

When the rosette is attached to a body under load the readings of the gauges show strains of: gauge 1 = 0·0001; gauge 2 = 0·0002, gauge 3 = 0·0003. Find by calculation the magnitude of the principal strains and their angles relative to gauge 1.

[e_1 = 0·0003 in direction of gauge 3. e_2 = 0·0001 in direction of gauge 1.]

6.13. Repeat problem 6.11 using Mohr's strain circle.

6.14. A rosette of three strain gauges on the surface of a metal plate under stress gave the following strain readings:

No. 1. at 0° +0·000592,
No. 2. at 45° +0·000308,
No. 3. at 90° −0·000432.

Find the magnitude of the principal strains and their directions relative to the axis of gauge 1. If E = 207 GN/m² and v = 1/3, find the principal stresses. [London Univ., B.Sc. 2]

[e_1 = 0·00064. e_2 = −0·00047 at 12° and 102° respectively counter-clockwise from gauge 1. f_1 = 112·5 MN/m². f_2 = −60 MN/m².]

6.15. A 45° strain gauge rosette measures strains of 0·000232, 0·000123, and −0·000080. The rosette is fitted to a steel member for which E = 207 GN/m² and v = 0·3. Draw Mohr's strain circle and find (a) the principal stresses, and (b) the angle to the nearest degree which the principal planes make with the line of the first gauge.

[f_1 = 48·58 MN/m². f_2 = −3·41 MN/m². Normal to first principal plane is 9° counter-clockwise. Normal to second principal plane is 99° counter-clockwise to line of gauge 1.]

6.16. Three strain gauges fixed to form a 60° rosette on a steel member all read strains of +0·0001. What two-dimensional stress system is acting at that point? (E = 200 GN/m². v = 0·25.)

[Equal tensile stresses acting in directions perpendicular to one another of magnitude 26·67 MN/m².]

6.17. Two strain gauges are attached to the surface of a horizontal shaft so that their lines of action are at 45° to the shaft axis and at 90° to one another. A recording apparatus is arranged so that the strain can be read while the shaft rotates. What will be the reading on each gauge (a) when the gauges are at the top, and (b) when the gauges are at the bottom, if the shaft is 100 mm diameter and is carrying a uniform torque of 4000 Nm? ($G = 80 \text{ GN/m}^2$, $E = 200 \text{ GN/m}^2$, $\nu = 0.3$.)

[$+0.000013$ and -0.000013 in both situations—readings will be constant.]

6.18. Two strain gauges are attached, one to the top surface and the other to the bottom surface, parallel with the axis of a uniform beam which bends under a uniform hogging bending moment. The top gauge reads 0·0001 and the bottom gauge -0.00005. If the beam is of steel ($E = 200 \text{ GN/m}^2$) and 15 mm deep, find (a) the position of the neutral axis, and (b) the maximum tensile and compressive stresses in the beam.

[(a) 5 mm above lower face. (b) $+20 \text{ MN/m}^2$, -10 MN/m^2.]

6.19. A solid horizontal shaft of 100 mm diameter carries a uniform torque of 2000 N m. It is supported by two self-aligning bearings 2 m apart so that it may be considered as a simply supported beam. The shaft weighs 605 N/m length. Calculate the bending and shear stresses at the bottom surface of the shaft at mid-span and (a) sketch an element of the shaft showing the stresses, (b) draw Mohr's stress circle, find the principal stresses, and indicate the position of the principal planes on a sketch of the shaft.

[Answer given in Fig. 6.8.]

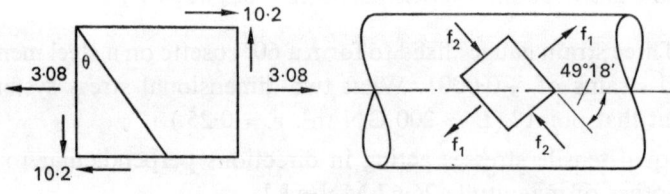

Fig. 6.8.

(*Note* that the direction of the shear stresses can be chosen to be in the reverse direction, the principal stress diagram would then be a mirror-image, about the centre line, of that given.)

6.20. For the shaft in problem 6.19, find the values of the principal strains and draw Mohr's strain circle. ($E = 200$ GN/m², $v = 0.3$.)

[Answer given in Fig. 6.9.]

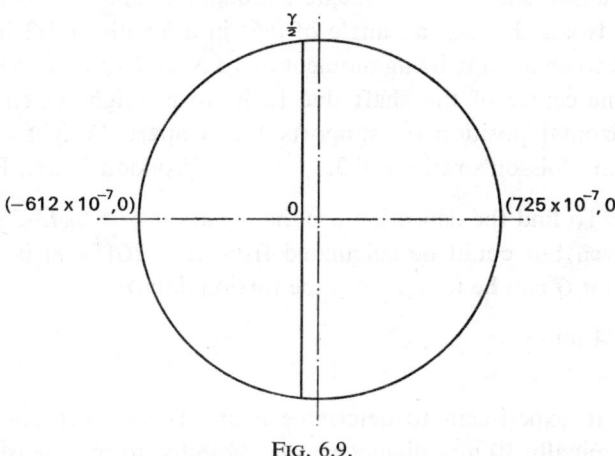

FIG. 6.9.

6.21. If two strain gauges whose lines of action are 90° apart and at 45° to the shaft axis were fastened to the mid-point of the shaft described in problems 6.19 and 6.20, what strain would the gauges read as the shaft rotated (a) when they were at the lowest point, and (b) when they were at the highest point?

[(a) 7.17×10^{-5}; -6.16×10^{-5}. (b) -7.17×10^{-5}; 6.16×10^{-5}.]

6.22. A bar of steel is loaded so that at a point on the surface there is a tensile stress parallel to the bar axis of 200 MN/m² and a shear stress of 100 MN/m². Find the principal stresses and from them calculate principal strains. Draw Mohr's strain circle and from it find the maximum shear strain. Hence calculate the maximum shear stress and compare it with the value calculated from $q_M = \frac{1}{2}(f_1 - f_2)$. ($E = 200$ GN/m². $G = 80$ GN/m². $\nu = 0.25$.)

[$f_1 = 241.4$ MN/m². $f_2 = -41.4$ MN/m².

$e_1 = 0.00126$. $e_2 = -0.000509$.

$\gamma_M = 0.00177$. $q_M = 141.4$ MN/m².]

6.23. A hollow shaft of 60 mm external diameter and 50 mm internal diameter twists through an angle of 0·6° in a length of 1·2 m when subjected to an axial twisting moment of 1 kN m. Estimate the deflection of the centre of the shaft due to its own weight when placed in a horizontal position on supports 1·2 m apart. Weight of shaft 178·7 N/m. Poisson's ratio = 0·3.   [London Univ., B.Sc. 2]

(*Hints.* To find the deflection use the equation $\delta = 5wL^4/384EI$. E is not given but could be calculated from $E = 2G(1+\nu)$ if G were known. But G can be found from the torsion data.)

[0·324 mm.]

6.24. In an experiment to determine E and G for steel, round test pieces nominally 10 mm diameter were subjected to pure tension and pure torsion tests. If the actual diameters were 9·92 mm but in calculations they were taken to be 10 mm, find the percentage error in the values of ν obtained from the relation between E, G, and ν assuming the correct values for E and G were 207 MN/m² and 80·5 MN/m² respectively.   [London Univ., B.Sc. 2]

[7 per cent.]

6.25. A block of steel carries a compressive stress of 90 MN/m² on the x face, a tensile stress of 30 MN/m² on the y face, and a stress f_z on the z face which just prevents all strain in that direction. If $E = 202.5$ GN/m² and $\nu = 0.3$, find e_x, e_y, and f_z.

[$e_x = -0.000462$. $e_y = +0.000308$. $f_z = -18$ MN/m².]

6.26. A rectangular block sustains stresses in three directions at right angles to each other of 75 MN/m² tensile, 60 MN/m² compressive, and 90 MN/m² tensile respectively. If $v = 0.286$ and $E = 195$ GN/m², calculate the strain in each direction and the torsion and bulk moduli.

[London Univ., B.Sc. 2]

[$e_x = 0.00034$. $e_y = -0.000548$, $e_z = +0.00044$. $K = 152$ GN/m². $G = 75.9$ GN/m².]

CHAPTER 7

Beam Deflections

Definitions and Theory

(a) Each of the following methods are worthy of study since certain systems can be analysed much more easily by one particular method than by the others. Strain energy methods are dealt with separately in Chapter 8.

(b) *Deflections by integration.* The deflection y, above horizontal, of a point on a beam located at x from the left-hand end of the beam is obtained from the equation

$$\frac{d^2y}{dx^2} = \frac{M}{EI},$$

where M is the bending moment at x, EI is the flexural rigidity of the beam, and the positive directions of y and x relative to the origin are

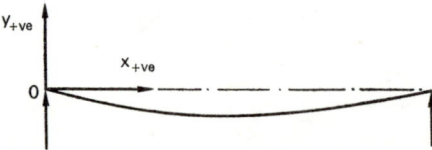

FIG. 7.1. Co-ordinate axes for beam deflections.

as shown in Fig. 7.1. (Some texts take the positive direction of deflection as downwards and use the equation $d^2y/dx^2 = -M/EI$.)

Integrating gives

$$\frac{dy}{dx} = \int \frac{M}{EI} dx + A,$$

where dy/dx represents the slope of the beam and

$$y = \iint \frac{M}{EI}\, dx\, dx + Ax + B.$$

To carry out the integration algebraically, M must be a continuous function of x from the left-hand end; to evaluate the constants of integration, values of y or dy/dx at corresponding values of x, must be known so that these values (boundary conditions) can be substituted to form two simultaneous equations so that A and B can be evaluated.

(c) *Macaulay's method*. If M is not a continuous function over the whole beam and if two boundary conditions cannot be obtained in the section over which it is a continuous function, then Macaulay's method may be used. Referring to Fig. 7.2, showing a beam carrying

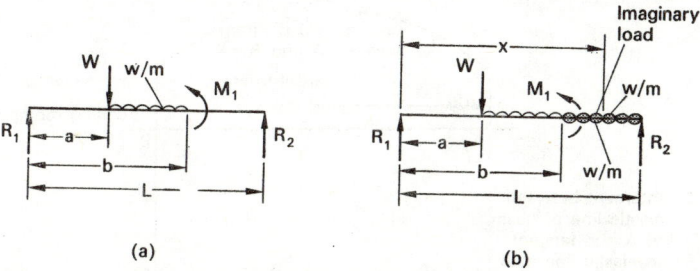

FIG. 7.2. Macaulay's method. (a) Actual loading of beam. (b) Imagined loading and location of section.

a point load, a uniformly distributed load and a moment applied at a point, the value of M is written at x, where x is such that it locates a section to the right of all applied loads and moments. The uniformly distributed load must be extended to the right-hand end of the beam by adding an imaginary downward load of intensity equal to the actual load, and this must be balanced by an equal and opposite upward load over the same length. For this beam

$$M = R_1 x - W[x-a] - \frac{w}{2}[x-a]^2 + \frac{w}{2}[x-b]^2 + M_1[x-b]^0.$$

The square brackets are integrated as a whole and are only removed after values of x have been substituted when finding the constants of integration (A and B) or the slope dy/dx or the deflection y; if any square bracket is found to give a negative value it is put equal to zero and the whole term discarded; the square bracket multiplying the moment M_1 is raised to the power 0 so that it equals 1 for evaluating M for $x > b$.

(d) *Method of superposition*. When several different types of load are applied simultaneously (as in Fig. 7.2), provided the combined loading does not cause plastic deformation, the slope or deflection at any point is the sum of the slopes or deflections which the beam would have due to each of the applied loads acting alone.

(e) *Moment-area method* (Fig. 7.3)

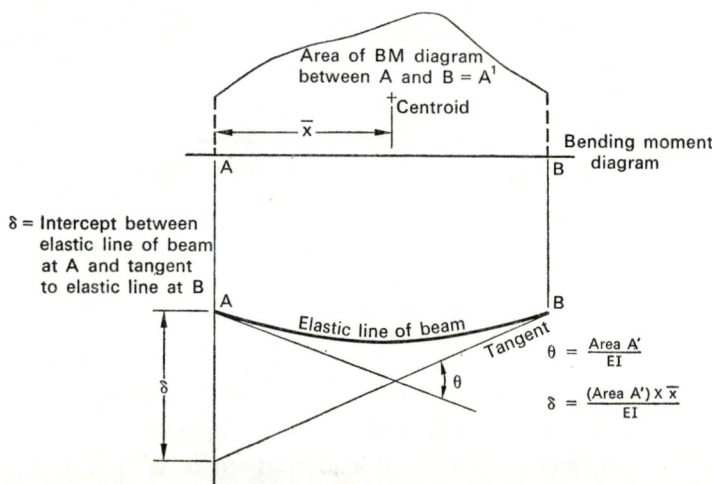

FIG. 7.3. Moment-area method of beam deflection.

Rule 1: *The change of slope (i.e. the angle between two tangents to the elastic line) between two points along the length of a loaded beam is equal to the area of the part of the bending moment diagram, between these two points, divided by EI.*

BEAM DEFLECTIONS

Rule 2: *The intercept on a vertical line drawn through point A on the beam, between the elastic line of the beam and the line which is a tangent to the elastic line at another point B, is equal to the first moment of area about A of the portion of the bending moment diagram between A and B, divided by EI.*

(f) *Fixed beams.* A beam firmly built-in at both ends is doubly redundant and four boundary conditions $y = 0$ at $x = 0$ and $x = L$, and $dy/dx = 0$ at $x = 0$ and $x = L$ will allow evaluation of the two constants of integration and two of the wall reactions if Macaulay's method is used.

A more simple approach uses superposition and moment-area (as in example 7.6) to show that if A_1 is the area of the bending-moment diagram of a simply supported beam of the same span and with the same loading as the fixed beam, and if A_2 is the area of the bending moment diagram which is due to the building-in of the ends

$$A_1 = A_2,$$
$$A_1 \bar{x}_1 = A_2 \bar{x}_2,$$

and, therefore, $\bar{x}_1 = \bar{x}_2$. It can be shown that the wall fixing moments M_A and M_B are

$$M_A = \frac{4A_1}{L} - \frac{6A_1\bar{x}_1}{L^2},$$

$$M_B = \frac{6A_1\bar{x}_1}{L^2} - \frac{2A_1}{L}.$$

Worked Examples

7.1. Derive from first principles, by integrating the elastic line, the formula for the central deflection of a simply-supported beam of span L, flexural rigidity EI carrying a central point load W. Explain why this method applies to a similar beam with a uniformly distributed load over the whole span but will not apply to a non-central point load.

Solution

Considering a section distance x from left-hand support. If x is taken to be less than $L/2$,

$$M = \frac{Wx}{2}.$$

$$\therefore EI\frac{d^2y}{dx^2} = \frac{Wx}{2}.$$

Integrating w.r.t. x,

$$EI\frac{dy}{dx} = \frac{Wx^2}{4} + A$$

and $\quad EIy = \dfrac{Wx^3}{12} + Ax + B.$

For a central point load it is clear that the beam will deflect symmetrically about its vertical centre line, and therefore the slope is zero at $x = L/2$.

$\therefore$ Boundary conditions are $y = 0$ at $x = 0$

$$\frac{dy}{dx} = 0 \text{ at } x = \frac{L}{2}.$$

Slope equation at $x = L/2$ is $0 = \dfrac{WL^2}{16} + A.$

$$\therefore A = -\frac{WL^2}{16}.$$

Deflection equation at $x = 0$ is $0 = 0 + 0 + B.$

$$\therefore B = 0.$$

Deflection equation from $x = 0$ to $x = L/2$ is

$$y = \frac{1}{EI}\left(\frac{Wx^3}{12} - \frac{WL^2 x}{16}\right)$$

at $x = L/2 \qquad y = -\dfrac{WL^2}{48EI}.$

BEAM DEFLECTIONS

In this problem, the bending-moment equation $M = Wx/2$ only applies if $x \not> L/2$. If $x > L/2$, then $M = (-W/2)(x-L)$.

Since the first equation was used, only values of x up to $-L/2$ can be used to find A and B.

If the point load was not in the centre, only one boundary condition $y = 0$ at $x = 0$ could be used in the section to the left of the point load since the value of x which makes $dy/dx = 0$ would be unknown; Macaulay's method would be appropriate for the non-central load situation. For a uniformly distributed load over the whole span,

$$M = \frac{wL}{2} - \frac{wx^2}{2}.$$

This applies over the whole span and the beam is symmetrical about its centre so that there are three boundary conditions available, i.e.

$$y = 0 \quad \text{at} \quad x = 0 \quad \text{and} \quad x = L,$$
$$\frac{dy}{dx} = 0 \quad \text{at} \quad x = \frac{L}{2},$$

any two of which can be used.

7.2. Derive the formula giving the maximum deflection of a beam of uniform section uniformly loaded over its whole length and simply supported at its ends.

If such a beam is a symmetrical I-section of a material whose modulus is 185 GN/m² and in which the maximum stress due to bending is limited to 125 MN/m², show that the deflection δ may be written in the form $\delta = KL^2/d$, where L is the span, d the overall depth. Also find the value of K.

Solution

If beam loading is w N/m,
Reaction at left-hand end $= wL/2$.

Moment at x from left-hand end $= \dfrac{wLx}{2} - \dfrac{wx^2}{2}.$

$$\therefore EI\frac{d^2y}{dx^2} = \frac{wLx}{2} - \frac{wx^2}{2},$$

$$EI\frac{dy}{dx} = \frac{wLx^2}{4} - \frac{wx^3}{6} + A,$$

$$EIy = \frac{wLx^3}{12} - \frac{wx^4}{24} + Ax + B$$

at $x = 0$, $y = 0$. $\therefore B = 0$.

$x = L$, $y = 0$. $\therefore A = \dfrac{wL^3}{24} - \dfrac{wL^3}{12} = -\dfrac{wL^3}{24}$.

Maximum deflection is at mid-span since the loading and beam are symmetrical about this point.

$$\therefore \delta = \frac{1}{EI}\left(\frac{wL^4}{96} - \frac{wL^4}{384} - \frac{wL^4}{48}\right)$$

$$= -\frac{5}{384}\frac{wL^4}{EI}.$$

Maximum bending stress $= My/I$,

$$125 \times 10^6 = \frac{wL^2 d}{16I},$$

$$\frac{w}{I} = \frac{125 \times 10^6 \times 16}{L^2 d}.$$

Substituting in the expression for y gives

$$\delta = -\frac{1000 \times 10^7}{7 \cdot 10 \times 10^{13}} \frac{L^2}{d},$$

$$\delta = -141 \times 10^{-6}\left(\frac{L^2}{d}\right),$$

which is of the required form with $K = -141 \times 10^{-6}$.

7.3. A simply supported beam is subjected to the loading shown in Fig. 7.4a. Calculate the deflection at a section 2·3 m from the left-hand end. ($E = 70$ GN/m^2 and $I = 8 \times 10^{-6}$ m^4.)

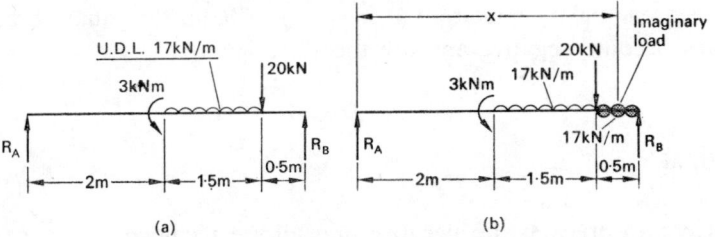

FIG. 7.4. (a) Actual loading. (b) Imagined loading and location of section.

Solution

Referring to Fig. 7.4b, the bending-moment equation must be written for the part of the beam between 3·5 m and 4 m from the left-hand support. Imaginary uniformly distributed loads of 17 kN/m positive and negative must be placed over this section.

To find reaction at A, take moments about B (note that imaginary loads are also included in this),

$$0 = (4R_A) - (3 \times 10^3) - (17 \times 10^3 \times 2 \times 1)$$
$$+ \left(17 \times 0.5 \times \frac{0.5}{2} \times 10^3\right) - (20 \times 0.5 \times 10^3).$$
$$\therefore R_A \doteqdot 11.22 \times 10^3 \text{ N}.$$

$$M = (11.22 \times 10^3 x) - (3 \times 10^3 [x-2]^0) - \left(\frac{17 \times 10^3}{2}[x-2]^2\right)$$
$$- (20[x-3.5] \times 10^3) + \left(\frac{17}{2}[x-3.5]^2 \times 10^3\right),$$

$$EI \frac{d^2 y}{dx^2} = M,$$

$$\frac{EI}{10^3} \frac{dy}{dx} = \frac{11.22 x^2}{2} - 3[x-2] - \frac{8.5}{3}[x-2]^3$$
$$- \frac{20}{2}[x-3.5]^2 + \frac{8.5}{3}[x-3.5]^3 + C,$$

$$\frac{EI}{10^3} y = \frac{11.22 x^3}{6} - \frac{3}{2}[x-2]^2 - \frac{8.5}{12}[x-2]^4$$
$$- \frac{20}{6}[x-3.5]^3 + \frac{8.5}{12}[x-3.5]^4 + Cx + D.$$

Using boundary condition at $x = 0$, $y = 0$, all the square bracket terms become negative and are therefore ignored.

$$\therefore D = 0$$

Also at $x = 4$, $y = 0$.

Here no brackets are negative and all are included:

$$0 = \frac{11.22 \times 64}{6} - 6 - \frac{8.5 \times 16}{12} - \frac{20}{48} + \frac{8.5}{12 \times 16} - 4C.$$

$$\therefore C = -25.5.$$

To find the deflection at $x = 2.3$, note that two square brackets are negative and are dropped.

$$y = \frac{10^3}{EI}\left[\frac{11.22}{6} \times 2.3^3 - \frac{3}{2}(0.3)^2 - \frac{8.5}{12}(0.3)^4 - 25.5 \times 2.3\right]$$

$$= \frac{10^3}{EI}[58.741 - 22.8]$$

$$= -64.2 \times 10^{-3} \text{ m when values of } E \text{ and } I \text{ are substituted.}$$

$$\therefore y = -64.2 \text{ mm}.$$

7.4. A beam of uniform section is built into solid walls at each end and has a clear span of 6 m. It carries a uniformly distributed load of 25 kN/m on the left-hand half and a point load of 120 kN at 4·5 m from the left-hand end. Find the force and moment reactions at the walls and draw a bending moment diagram for the beam, inserting principal values. [London Univ.]

(The following solution uses Macaulay's method. For the analysis of fixed beams the moment-area and superposition methods given later are often far easier. Many people, however, prefer Macaulay since it provides a routine method that can be employed without much thought.)

Solution

Refer to Fig. 7.5.

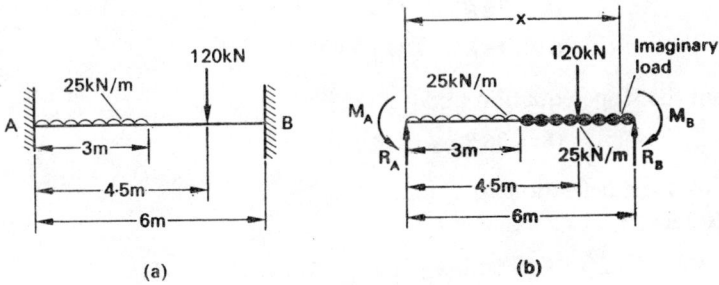

FIG. 7.5. (a) Actual loading. (b) Imagined loading and location of section.

The value of x must be taken to be between 4·5 m and 6 m from the left-hand end. Imaginary distributed loads of 25 kN/m must be applied to the beam at the top and bottom over the right-hand half. Then, taking 10^3 from the right-hand side as a common factor,

$$10^{-3}M = R_A x + \frac{25}{2}[x-3]^2 - \frac{25x^2}{2} - 120[x-4\cdot5] - M_A, \quad (1)$$

$$\frac{EI}{10^3} \cdot \frac{d^2y}{dx^2} = R_A x + \frac{25}{2}[x-3]^2 - \frac{25x^2}{2} - 120[x-4\cdot5] - M_A,$$

$$\frac{EI}{10^3} \cdot \frac{dy}{dx} = \frac{R_A x^2}{2} + \frac{25}{6}[x-3]^3 - \frac{25x^3}{6} - \frac{120}{2}[x-4\cdot5]^2 - M_A x + A,$$

$$\frac{EIy}{10^3} = \frac{R_A x^3}{6} + \frac{25}{24}[x-3]^4 - \frac{25x^4}{24} - \frac{120}{6}[x-4\cdot5]^3 - \frac{M_A x^2}{2}$$
$$+ Ax + B.$$

Boundary conditions are $dy/dx = 0$ at $x = 0$ and $x = 6$,

$y = 0$ at $x = 0$ and $x = 6$.

$$\therefore A = B = 0.$$

The boundary conditions remaining, i.e. $dy/dx = 0$, $y = 0$ at $x = 6$, can now be used to form two more equations that are solved to give values for the unknowns R_A and M_A,

i.e.
$$0 = 18R_A + 112\cdot5 - 900 - 135 - 6M_A,$$
$$0 = 18R_A - 6M_A - 922\cdot5$$

from the slope equation and
$$0 = 36R_A - 1333 - 18M_A$$

from the deflection equation.

Solving simultaneously,
$$R_A = 79\cdot66 \text{ kN} \quad \text{and} \quad M_A = 85\cdot25 \text{ kN m}.$$

The remaining two unknowns can be found by statical methods,

i.e. sum of vertical forces = 0 (positive upwards),
$$R_A + R_B - 75 \times 10^3 - 120 \times 10^3 = 0,$$
$$R_B = (120 - 4\cdot66) \times 10^3 \text{ N}$$
$$= 115\cdot33 \text{ kN}.$$

Sum of moments about $A = 0$ (clockwise positive),
$$-M_A \times 10^3 + 25 \times 10^3 \times \tfrac{9}{2} + 120 \times 10^3 \times \tfrac{9}{2} - R_B \times 10^3 \times 6 + M_B \times 10^3 = 0$$
$$-85\cdot25 + 112\cdot5 + 540 - 691\cdot98 + M_B = 0,$$
$$M_B = 124\cdot8 \text{ kN m}.$$

Reactions at A are 79·66 kN and 85·25 kN m and at B are 115·33 kN and 124·8 kN m.

The values for the bending-moment diagram can be obtained by substituting values in the first equation for M [i.e. eqn. (1)]:

at $x = 0$, $\quad M = -10^3 M_A,$
$$= -85\cdot25 \text{ kN m};$$
at $x = 3$, $\quad 10^{-3}M = 239 - 112\cdot5 - M_A$
$$= 239 - 197\cdot75$$
$$= 41\cdot25,$$
$$M = 41\cdot25 \text{ kN m}.$$

BEAM DEFLECTIONS 161

M changes sign between these two points; thus it must be zero somewhere between. To find the location of this point put $M = 0$ and solve for x, i.e. when $x = 1\cdot443$ m:

at $x = 6$, $\qquad M = -10^3 M_B$
$\qquad\qquad\qquad = -124\cdot8$ kN m.

To find M at $x = 4\cdot5$ it is easiest to work from first principles from the right-hand end, then at $x = 4\cdot5$. $M = 48\cdot2$ kN m and there is another point of contraflexure at $1\cdot08$ m from right-hand support. The resulting diagram is shown in Fig. 7.6.

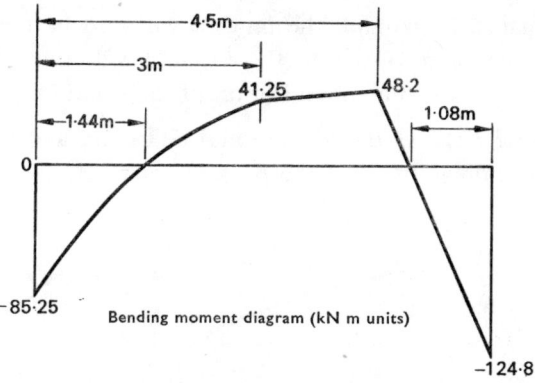

FIG. 7.6. Bending-moment diagram.

7.5. A cantilever of uniform section is built into a rigid wall so that it protrudes horizontally a distance L. It is propped at the free end and carries a load W at a distance d from the fixed end.

(a) If the prop is removed, show that the deflection of the free end is

$$\frac{Wd^2}{6EI}(3L-d).$$

(b) If the prop applies a force P to the free end such that the free end does not deflect at all, show that

$$P = \frac{Wd^2}{2L^3}(3L-d).$$

(c) For the particular case $L = 4875$ mm, $d = 3625$ mm, and $W = 80$ kN, draw the bending-moment diagram and show on it the magnitudes of maximum positive and negative bending moments. [London Univ.]

Solution

The moment-area rule 2 states that the moment of the area of the bending-moment diagram divided by EI gives the intercept between the elastic line and a tangent. For cantilevers and fixed beams the point of tangency can be taken at a built-in end. If the built-in end is, and remains, horizontal, the tangent will also be the horizontal datum and the intercept will be the required deflection of the elastic line. This fact leads to a simplification of the solution.

(a) Figure 7.7 shows the loading and deflection and the bending-moment diagram when the prop is not in place.

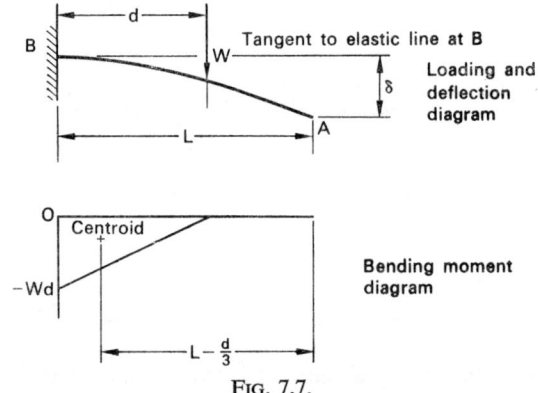

FIG. 7.7.

Moment of area of bending-moment diagram about point A is

$$\frac{d}{2} \times Wd \times (L-d/3) = \frac{Wd^2}{2}(L-d/3).$$

By rule 2, $$\delta = \frac{Wd^2}{6EI}(3L-d),$$

which is the required deflection.

BEAM DEFLECTIONS

(b) Using the superposition principle, the deflection of the free end, due to the load W and the force P in the prop, will be the sum of the deflections each would produce if it acted alone. As before, the deflection will be $1/EI \times$ (first moment of area of the bending-moment diagram taken about the free end). In fact the total deflection is stated to be zero.

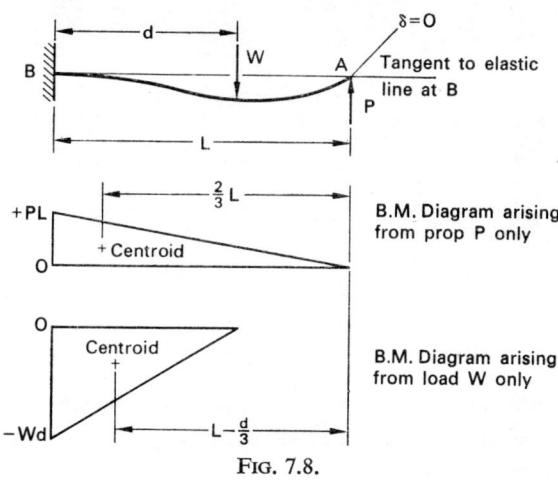

FIG. 7.8.

∴ First moment of area of the bending-moment diagram = 0.
$1/EI \times$ moment of prop bending-moment diagram is seen from Fig. 7.8 to be

$$\frac{2PL^3}{6EI}.$$

Also from Fig. 7.8,

$\frac{1}{EI} \times$ moment of loading bending-moment diagram

$$= \frac{Wd^2}{6EI}(3L-d).$$

∴ $\frac{2PL}{6EI} + \frac{Wd^2}{6EI}(3L-d) = 0,$

$$P = -\frac{Wd^2}{2L^3}(3L-d),$$

the negative sign indicating that P and W act in opposite directions.

(c) Substituting values in

$$P = -\frac{Wd^2}{2L^3}(3L-d)$$

gives $P = 50$ kN upwards.

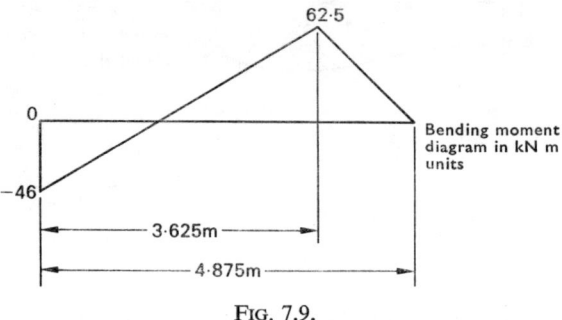

FIG. 7.9.

The total bending-moment diagram shown in Fig. 7.9 is obtained by adding the two triangular diagrams shown in Fig. 7.8 with appropriate values.

7.6. A horizontal beam has a span L and is firmly built-in at each end. A point load W acts such that its distances from the left and right walls are a and b respectively. If the fixing moments are M_A at the left and M_B at the right, show that

$$M_A = \frac{b}{L}M_f, \quad M_B = \frac{a}{L}M_f,$$

where $M_f = (Wab)/L$ and is the maximum bending moment of the equivalent simply supported beam.

Solution

Figure 7.10 shows at (a) the loading and reactions on the actual beam, at (b) the loading and reactions on the equivalent simply supported

BEAM DEFLECTIONS 165

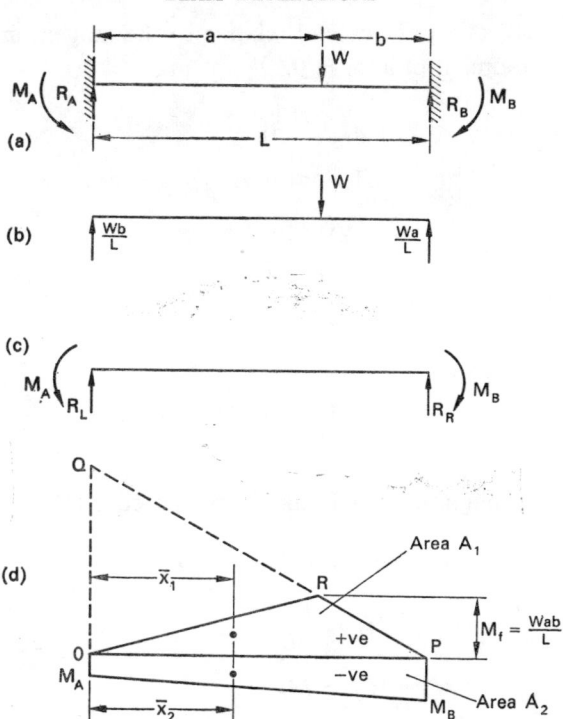

FIG. 7.10. (a) Actual loading. (b) Equivalent free beam. (c) Fixing reactions. (d) Bending-moment diagram.

beam, and at (c) the remaining moment and force reactions due to the beam being built-in. The superposition principle indicates that since the total loading (a) can be obtained by combining the loadings (b) and (c), the total bending moment and the total deflection can be obtained by combining bending moments and deflections.

The bending-moment diagrams are shown at (d) with the free diagram above the axis since it is positive, and the **fixing diagram below** since it is negative.

Since the change of slope of one end relative to the other is zero, total area = 0.

$$A_1 + A_2 = 0,$$
$$A_1 = -A_2. \tag{1}$$

Since the deflection of one end relative to the tangent at the other is zero total, moment of area = 0.

$$A_1\bar{x}_1 + A_2\bar{x}_2 = 0,$$

$$A_1\bar{x}_1 = -A_2\bar{x}_2,$$

and since
$$A_1 = -A_2,$$

$$\bar{x}_1 = \bar{x}_2. \qquad (2)$$

$$A_1 = \frac{L}{2}\frac{Wab}{L},$$

$$A_2 = \frac{L}{2}(M_A + M_B).$$

Considering magnitudes only and applying eqn. (1),

$$Wab = (M_A + M_B)L,$$

or using
$$M_f = \frac{Wab}{L},$$

$$M_A + M_B = M_f.$$

To locate the centroids it is best to divide the actual figures into parts whose centroids can be found by inspection or are given in standard tables. Thus the trapezium is a rectangle plus a triangle.

$$\bar{x}_2 = \frac{M_A L \times L/2 + L/2(M_B - M_A) \times 2L/3}{L/2(M_A + M_B)}$$

$$= \frac{L/3(3M_A - 2M_A + 2M_B)}{(M_A + M_B)},$$

$$\bar{x}_2 = \frac{(M_A + 2M_B)L}{3(M_A + M_B)}.$$

The area A_1 can be considered as triangle OPQ (whose centroid is $L/3$ from O) less triangle ORQ (whose centroid is $a/3$ from O).

Now
$$OQ = \frac{L}{b} M_f,$$

$$\bar{x}_1 = \frac{L/2(M_fL/b) L/3 - a/2(M_fL/b) a/3}{M_f L/2},$$

$$\bar{x}_1 = \frac{L^2}{3b} - \frac{a^2}{3b}$$

$$= \frac{(L+a)(L-a)}{3b}.$$

But
$$b = L - a.$$

$$\therefore \bar{x}_1 = \frac{L+a}{3},$$

and since
$$\bar{x}_1 = \bar{x}_2,$$

$$\frac{(M_A + 2M_B)L}{3(M_A + M_B)} = \frac{L+a}{3}.$$

$$\therefore M_A = M_B \frac{(L-a)}{a},$$

$$M_A = M_B \frac{b}{a},$$

$$M_B \frac{b}{a} + M_B = M_f,$$

$$M_B(a+b) = aM_f,$$

$$(a+b) = L.$$

$$\therefore M_B = \frac{a}{L} M_f.$$

$$M_A = \frac{b}{L} M_f.$$

7.7. A steel girder 0·5 m deep, symmetrical about both principal axes and $I_{xx} = 8\cdot541 \times 10^{-4}$ m⁴, has a span of 10 m and is rigidly built-in at both ends. The loading, including self-weight, consists of a uniformly distributed load of 25 kN/m on the whole span and **three** equal loads W at the centre and two quarter points. Find the magnitude of W if the maximum bending stress is 125 MN/m². ($E = 200$ GN/m².)

[London Univ.]

Solution

The bending-moment diagram will be the result of combining the three diagrams shown in Fig. 7.11: (a) the fixing diagram, (b) the diagram due to the point loads on a simply-supported beam, and (c) the diagram due to the distributed load on a simply-supported beam.

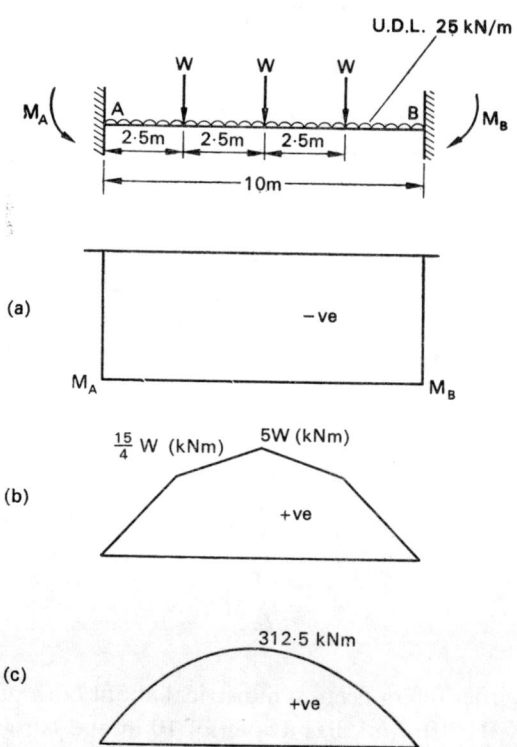

FIG. 7.11. (a) Fixing diagram. (b) Point loads on a simply-supported beam. (c) Distributed load on a simply-supported beam.

But moment-area considerations lead to the conclusion that the total area of bending-moment diagram = 0.

Therefore adding the areas of the three diagrams (remembering that the fixing area is negative) and equating to zero gives

$$0 = -10M_A + 2\left(\frac{2 \cdot 5}{2} \times \frac{15W}{4} + \frac{2 \cdot 5}{2} \times \left[\frac{15W}{4} + 5W\right]\right) + \frac{2 \times 10 \times 312 \cdot 5}{3},$$

$M_A = 3 \cdot 125W + 208 \cdot 33$ kN m.

This is the maximum bending moment on the beam and maximum stress due to bending is

$$\frac{(3 \cdot 125W + 208 \cdot 33)\, 0 \cdot 25}{8 \cdot 541 \times 10^{-4}} \text{ kN/m}^2.$$

But the maximum stress is to be 125×10^3 kN/m².

$$\therefore \frac{(3 \cdot 125W + 208 \cdot 33)\, 0 \cdot 25}{8 \cdot 541 \times 10^{-4}} = 125 \times 10^3,$$

$$W = 70 \text{ kN}.$$

Problems

7.8. A cantilever protrudes horizontally a distance of 3 m from a vertical wall. It is made of a steel tube, outside diameter 150 mm, internal diameter 125 mm. ($E = 200$ GN/m².)

A vertical load of 10 kN is applied at a point 1·75 m from the wall. Find, without using the moment-area method, (a) the deflection of the point of application of the load, (b) the slope of the free end, and (c) the deflection of the free end.

[(a) 6·95 mm. (b) 0·00596 rad. (c) 14·4 mm.]

7.9. A horizontal cantilever, rigidly built into a wall, is made up of two identical U-section bars bolted rigidly together with special double ended bolts as shown in Fig. 7.12 so that the whole is equivalent to a single bar with second moment I. During construction work a

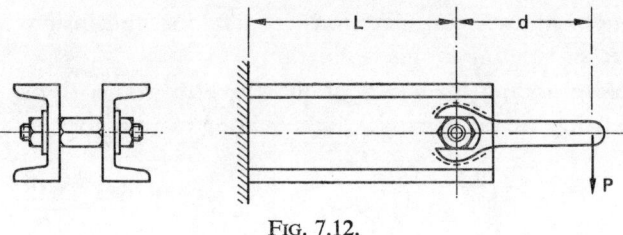

Fig. 7.12.

fitter puts a spanner of length d on the outermost central bolt which is at a distance L from the wall and pulls down with a force P. The spanner is horizontal and the man fails to turn the bolt.

Derive an expression for the deflection, δ below the horizontal, of the bolt on which the spanner fits assuming the modulus of elasticity of the material is E and that the whole deflection is due to bending.

$$\left[\delta = -\left(\frac{PL^3}{3} + \frac{PL^2 d}{2}\right) \cdot \right]$$

7.10. A rectangular section timber beam has a span of 5 m and is simply supported at the ends, it is required to support a total load of 45 kN uniformly distributed over the span. Find the breadth and depth of the section so that the maximum bending stress does not exceed 7 MN/m² and the maximum deflection does not exceed 9·5 mm. ($E = 10$ GN/m².) [London Univ.]

[Depth = 383·8 mm. Breadth = 163·6 mm.]

7.11. A horizontal beam is simply supported at two points 6 m apart and carries two vertical point loads each 40 kN symmetrically placed about the centre line of the span and 3·5 m apart. Calculate the maximum deflection and the slope of the beam at the supports if $E = 200$ GN/m² and $I_{NA} = 10^{-4}$ m⁴.

[− 10·6 mm. ±0·0059 rad.]

7.12. A horizontal beam of length 3·6 m has a pin support at the left-hand end and a simple support at 2·4 m to the right. A downward force of 500 N is applied at the overhanging end. If $I = 25 \times 10^{-8}$ m^4 and $E = 200$ GN/m^2, find the deflection of the overhanging end and the slope at the simple support.

[Deflection $= -17.28$ mm. Slope $= -0.0096$ rad.]

7.13. A rolled-steel beam of I-section is 375 mm deep and has $I_{NA} = 1.8 \times 10^{-4}$ m^4. The beam is simply supported at its ends and is required to carry a uniformly distributed load, including self-weight, of 25 kN/m over its whole length. Find the greatest allowable span to satisfy both the following conditions: (a) bending stress not to exceed 140 MN/m^2, and (b) maximum deflection not to exceed one four-hundredth of the span.

What will be the values of maximum stress and deflection for the actual span calculated to satisfy the conditions? ($E = 200$ GN/m^2.)

[London Univ.]

[6.515 m. $f_{max} = 138$ MN/m^2. $\delta = 16.3$ mm.]

7.14. A uniform cantilever is 4.25 m long, 0.5 m deep, and has a maximum bending stress of 75 MN/m^2 due to a load of 60 kN uniformly distributed along the length and a point load of 40 kN at the free end. Calculate the deflection of the free end. ($E = 200$ GN/m^2.)

[London Univ.]

[8.063 mm.]

7.15. A strip of brass 75 mm by 19 mm by 3000 mm weighs 340 N and is observed to have a slight curvature along its length. It is supported at its ends with the 75 mm dimension horizontal and the sag in the middle is observed to be 41.8 mm. When turned completely over, the sag is 28.8 mm. Find the sag due to the original curvature of the strip and the modulus of elasticity of the brass.

[London Univ.]

[35.3 mm. 79 GN/m^2.]

7.16. A simply-supported beam of uniform section and span L carries a point load W at mid-span. Derive an expression, in terms of W, L, E, and I for the deflection of the mid-span point.

If the beam described had $L = 4\cdot 25$ m, $W = 32\cdot 5$ kN, $I = 0\cdot 33 \times 10^{-4}$ m⁴, and a central deflection of $7\cdot 7$ mm, what would be the value of E for the material?

If the same total load were applied uniformly distributed over the whole span, what would be the deflection?

[$(WL^3)/(48EI)$. 205 GN/m². 4·8 mm.]

7.17. A beam of constant I-section is supported freely at each end and spans 5 m. It carries a concentrated load of 120 kN at 2 m from the left-hand support. $I_{NA} = 8\cdot 74 \times 10^{-5}$ m⁴. What is the deflection under the load if $E = 200$ GN/m²?

[16·5 mm.]

7.18. A uniform beam rests on two simple supports which are a distance L apart. One support is at the left-hand end of the beam, but the beam overhangs the right-hand support by a length $L/3$. A uniformly distributed load w per unit length covers a length of $L/3$ on each side of the right-hand support. If the flexural rigidity of the beam is EI, find an expression for the deflection of the right-hand end of the beam.

$$\left[\frac{-5wL^4}{1458EI}\cdot\right]$$

7.19. Explain why it is possible to solve problem 7.11 by integration of the elastic line without using Macaulay's method. List all the boundary conditions available for finding the constants of integration. What difference would it make to the method of solution if only the right hand of the two 40 kN loads were acting?

7.20. A horizontal beam AB is simply supported at each end over a span of 7·25 m and carries a uniformly distributed load of 15 kN/m including its own weight. A clockwise moment of 15 kN m acts about a horizontal axis perpendicular to the beam length at point C, 3 m from end A.

BEAM DEFLECTIONS 173

If $EI = 40$ MN m², calculate the slope and deflection of the beam at C. [London Univ.]

[−16·08 mm. −0·00204 rad.]

7.21. A beam of uniform section has a length of 6 m and is simply supported at each end. There are three point loads 90 kN at 1·5 m, 100 kN at 2·75 m, and 50 kN at 5 m from the left-hand end. Find the deflection at mid-span and the slope at each end if $I = 3·5 \times 10^{-4}$ m⁴ and $E = 200$ GN/m². [London Univ.]

[−11·9 mm. −0·0065 rad at left-hand end. +0·006 rad at right-hand end.]

7.22. A symmetrical I-section steel beam is 0·3 m deep and has $I_{NA} = 0·85 \times 10^{-4}$ m⁴. It is simply supported in a horizontal position at either end of its length of 5 m. There are three vertical loads; W at 1·25 m; 1·5 W at 2 m; 2W at 4 m from the left-hand end. Calculate W if the maximum stress must not exceed 120 MN/m². Find also the position and magnitude of maximum deflection if $E = 200$ GN/m². [London Univ.]

[$W = 20·3$ kN. Maximum deflection = 10·09 mm at 2·48 m from left-hand support.]

7.23. A uniform steel beam is 7 m long and rests symmetrically on two simple supports 5 m apart. The beam carries three point loads 20 kN at the left-hand end, 40 kN at the right-hand end, and 120 kN at 3 m from the left-hand end. If $EI = 37$ MN m², find the deflection at each load point and indicate whether it is above or below the horizontal. [London Univ.]

[At $x = 0, y = 3·206$ mm. At $x = 3, y = -5·405$ mm. At $x = 7, y = -1·924$ mm.]

7.24. A straight horizontal beam 3 m long has two simple supports 2 m apart; one support is at an end of the beam. A uniformly distributed load of 30 kN/m covers the whole length. $I_{NA} = 1·25 \times 10^{-4}$

m⁴, $E = 200$ GN/m². Find (a) the slope at each end, and (b) the deflection relative to the line of the supports of the point at mid span.

[London Univ.]

[Left-hand end slope $= -0.0002$ rad. Right-hand end slope $= -0.0002$ rad. Deflection $= -0.1$ mm.]

7.25. A simply supported beam of span L carries a uniformly distributed load of intensity w per unit length extending from one end over a length of $L/3$. Using Macaulay's method, obtain the coefficient K in the formula $\delta = (wL^4)/KEI$ for the deflection of the point of mid-span.

[$K = 311$.]

7.26. A beam 6 m long is simply supported at both ends and carries two point loads, one of 40 kN at 3 m from the left-hand support and one of 60 kN at 2 m from the right-hand support. If $E = 200$ GN/m² and $I = 5 \times 10^{-4}$ m⁴, find, using Macaulay's method, the deflection under each load and the position and magnitude of maximum deflection.

[4·1 mm under 40 kN. 3·668 mm under 60 kN. Maximum deflection $= 4.114$ mm.]

7.27. Using the moment-area method, find the deflection of the free end of a horizontal cantilever 1 m long due to a point load of 100 kN acting halfway along its length. ($E = 200$ GN/m² and $I = 5.21 \times 10^{-5}$ m⁴.)

[1 mm.]

7.28. Rework problem 7.8 using the moment-area method.

7.29. A horizontal cantilever of effective length $3a$ carries two concentrated loads: W at a distance a from the fixed end; W_1 at a distance a from the free end. Obtain a formula for the maximum deflection arising from this loading. [London Univ., part question]

$$\left[y = \frac{2a^3}{3EI}(2W + 7W_1). \right]$$

7.30. For the cantilever described in problem 7·29, find W and W_1 when $a = 1$ m, the symmetrical section is 250 mm deep with $I = 85 \times 10^{-6}$ m^4, the maximum deflection is 6·25 mm, the maximum bending stress is 90 MN/m^2, and $E = 200$ GN/m^2.

[36·58 kN. 12·307 kN.]

7.31. A horizontal pole 50 mm diameter is 6 m long, built-in at one end, and carries a load W at the free end. It is propped at the midpoint such that the deflection at the free end is zero relative to a horizontal line through the fixed end. Find the ratio of the load W to the prop force R.

[$W/R = 5/16$.]

7.32. Deduce a formula for the deflection of a beam of constant section when loaded at the centre and fixed horizontally at the ends. A timber beam has a rectangular section 0·125 m wide by 0·3 m deep, a span of 6 m, and is built in horizontally at each end. Find (a) the central point load that will produce a maximum bending stress of 7 MN/m^2, and (b) the deflection under this load. ($E = 12$ GN/m^2.)

[London Univ.]

[$(WL^3)/(192EI)$. 17·5 kN. 5·8 mm.]

7.33. A beam of uniform section, $I = 1·85 \times 10^{-4}$ m^4, spans 6 m and is fixed horizontally at each end. It carries a point load of 120 kN at 3·625 m from one end and may be assumed to be weightless; find (a) the fixing moments and reactive forces, and (b) the position and magnitude of maximum deflection. ($E = 200$ GN/m^2.)

[London Univ., part question]

[$M_A = 68·16$ kN m. $R_A = 41·52$ kN. $M_B = 103·9$ kN m. $R_B = 78·48$ kN. $y_{\max} = -3·313$ mm at 3·283 m from A.]

7.34. A rolled-steel joist 6 m clear span is built into walls at each end and supports two loads of 50 kN each at a distance of 2·125 m from a wall. Sketch the bending-moment and shear-force diagrams. Find a suitable section for the beam if depth is three times the breadth of the flanges and twelve times the thickness of web and of flanges. The maximum stress is to be 90 MN/m². [London Univ.]

[Figure 7.13 gives shear-force and bending-moment diagrams. Beam section $d = 289\cdot4$ mm, $b = 96\cdot5$ mm, and $24\cdot1$ mm thick.]

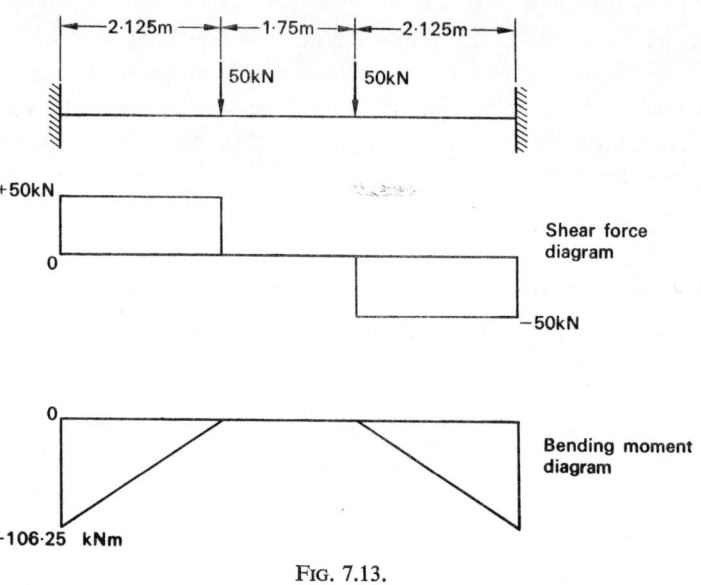

FIG. 7.13.

7.35. A horizontal I-section beam rigidly built in at the ends and 8 m long carries a uniformly distributed load of 100 kN as well as a central load of 40 kN. If the bending stress is limited to 75 MN/m² and the deflection must not exceed 2·5 mm, find the depth of section required. ($E = 200$ GN/m².)

[673·6 mm ($I = 4\cdot788 \times 10^{-4}$ m⁴).]

CHAPTER 8

Strain Energy Methods

Definitions and Theory

(a) The strain energy stored in a member being elastically deformed under various conditions of loading can be shown to be:

(1) Tension or compression,

$$U = \frac{f^2}{2E} \times \text{volume of the bar,}$$

where U is the strain energy or

$$U = \frac{P^2 L}{2AE},$$

where P is the force carried by the bar of length L and cross-section A.

(2) Shear,

$$U = \frac{q^2}{2G} \times \text{volume of the bar.}$$

(3) Bending,

$$U = \int_0^L \frac{M^2}{2EI} dx,$$

where M is the bending moment at a point distant x from one end of the beam and I is the second moment about the neutral axis.

(4) Torsion,
$$U = \frac{T^2 L}{2GJ},$$

where T is the torque and J the polar second moment.

(i) For a solid shaft,
$$U = \frac{q_{max}^2}{4G} \times \text{volume of the shaft}.$$

(ii) For a hollow shaft,
$$U = \frac{q_{max}^2}{4G}\left[\frac{D^2+d^2}{D^2}\right] \times \text{volume of the shaft}.$$

(b) **Castigliano's first theorem**: *If the total strain energy expressed in terms of the external loads be partially differentiated with respect to any one of the external loads, the result gives the displacement of that load along its line of action.*

This theorem may be used to obtain displacements of single structural members of frameworks. For torsion-carrying members the torsional displacement can be obtained by differentiating with respect to a moment. If the displacement is required in a direction in which no force is acting, an imaginary force may be applied at the required point which, after the calculation of strain energy and partial differentiation, is placed equal to zero.

(c) **Castigliano's second theorem**: *The partial differential coefficient of the total strain energy of a frame with respect to the load in a redundant member is equal to the initial lack of fit of that member.*

Worked Examples

8.1. Verify Castigliano's first theorem for (a) a vertical tension rod of uniform cross-sectional area A, modulus E, length L, carrying an axial tensile load P, and (b) a horizontal cylindrical shaft of diameter D, modulus G, length L, fixed at one end and carrying a torque T at the other.

Solution
Refer to Fig. 8.1.

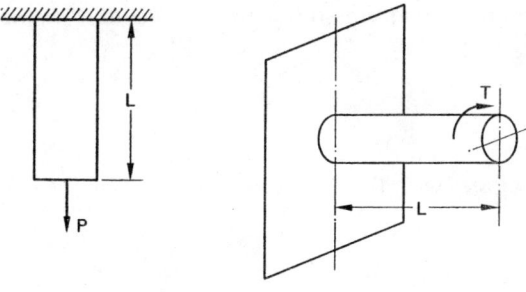

Fig. 8.1.

(a) The rod is under uniaxial tension.

$$\therefore E = \frac{\text{stress}}{\text{strain}}$$

$$= \frac{P/A}{x/L},$$

where x is the extension due to the load.

$$\therefore x = \frac{PL}{AE}.$$

Applying Castigliano's theorem, the strain energy U in the bar is $(P^2L)/2AE$ and the partial differential with respect to P, which is $(\partial U/\partial P) = (PL/AE)$ which is the extension of the point of application of P and agrees with the previous result.

(b) From simple torsion theory,

$$\frac{T}{J} = \frac{G\theta}{L}.$$

Therefore angular displacement of cross-section where torque is applied

$$= \theta = \frac{TL}{GJ}.$$

Strain energy stored in the bar $= U = \dfrac{T^2 L}{2GJ}$.

Castigliano's theorem states that the angular displacement

$$\theta = \frac{\partial U}{\partial T}, \quad \text{but} \quad \frac{\partial U}{\partial T} = \frac{TL}{GJ},$$

which agrees with the displacement found by the simple method.

8.2. A circular section solid bar has a length K cut from one end and welded at right angles to the remaining length L. The bar is then solidly fixed to a vertical wall so that the whole bar is horizontal and

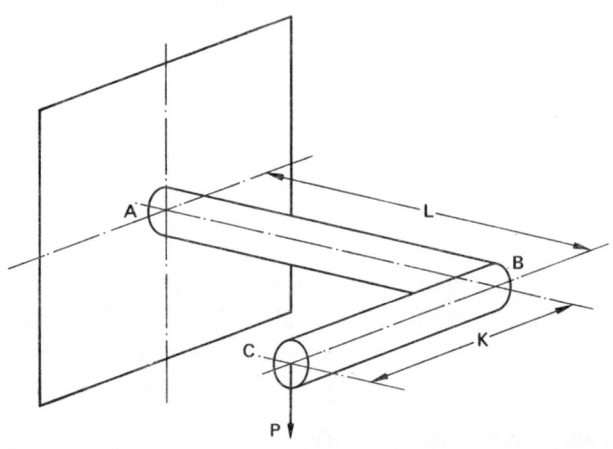

Fig. 8.2.

the end section is parallel to the wall (Fig. 8.2). Find the vertical deflection of point C due to the combination of (a) bending of section AB, (b) twisting of section AB, and (c) bending of section BC when a vertical load P is applied at C.

Solution

In section AB there will be a bending moment at x from B of Px.
Therefore strain energy due to bending,

$$U_1 = \int_0^L \frac{P^2 x^2}{2EI} \, dx$$

$$= \frac{P^2 L^3}{6EI}.$$

There will also be a torque PK.
Therefore strain energy due to torque,

$$U_2 = \frac{P^2 K^2 L}{2GJ}.$$

In section BC the strain energy due to bending,

$$U_3 = \int_0^K \frac{P^2 x^2}{2EI} \, dx$$

$$= \frac{P^2 K^3}{6EI}.$$

Total strain energy,

$$U = U_1 + U_2 + U_3$$

$$= \frac{P^2 L^3}{6EI} + \frac{P^2 K^2 L}{2GJ} + \frac{P^2 K^3}{6EI}.$$

Deflection of point C,

$$= \frac{\partial U}{\partial P}$$

$$= \frac{PL^3}{3EI} + \frac{PK^2 L}{GJ} + \frac{PK^3}{3EI}.$$

8.3. Find the deflection, using Castigliano's theorem, of a point at 1 m from the support of a simply-supported beam of 3 m span which has a second moment of area of 2×10^{-5} m^4 and carries a uniformly distributed load of 10 kN/m. ($E = 200$ GN/m^2.)

Solution

In order to find the deflection δ by Castigliano it is necessary to evaluate $\partial U/\partial P$, where P is a force acting at the point where, and in the line along which, the deflection is required. Since this is a uniformly loaded beam there is no point load at the desired location, and an imaginary load P ($=0$) must be assumed to act at the point where the deflection is required.

U could be evaluated from

$$\int_0^L \frac{M^2 \, dx}{2EI} \quad \text{and then} \quad \frac{\partial U}{\partial P}$$

found, all using the imaginary load P in addition to the actual load.

However,

$$\frac{\partial U}{\partial P} = \frac{\partial}{\partial P} \int_0^L \frac{M^2 \, dx}{2EI},$$

and since the differentiation is with respect to P while the integration is with respect to x, it is immaterial mathematically which is done first, i.e.

$$\frac{\partial U}{\partial P} = \int_0^L \frac{\partial}{\partial P}\left(\frac{M^2}{2EI}\right) dx$$

$$\frac{\partial U}{\partial P} = \frac{I}{EI} \int_0^L \frac{M \, \partial M}{\partial P} \, dx.$$

Using this form, M and $\partial M/\partial P$ can be found using the actual and the

STRAIN ENERGY METHODS 183

imaginary loads, and then P can be put equal to its real value, zero. This usually simplifies the expression to be integrated.

Referring to Fig. 8.3, the reaction

$$R_A = 15\,000 + \frac{2P}{3} \quad \text{while} \quad R_B = 15\,000 + \frac{P}{3}.$$

The equation for the bending moment is discontinuous over the length of the beam, and the integration must therefore be performed in two parts.

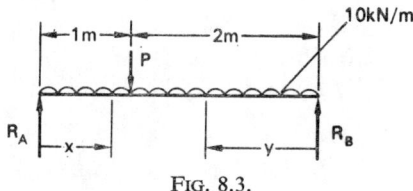

FIG. 8.3.

Over the left-hand section,

$$M_x = 15\,000x + \frac{2Px}{3} - \frac{10\,000x^2}{2}.$$

Over the right-hand section,

$$M_y = 15\,000y + \frac{Py}{3} - \frac{10\,000y^2}{2}.$$

$$U = \int_0^1 \frac{M_x^2\,dx}{2EI} + \int_0^2 \frac{M_y^2\,dy}{2EI},$$

$$\delta = \frac{\partial U}{\partial P} = \frac{1}{EI}\left[\int_0^1 M_x \frac{\partial M_x}{\partial P}\,dx + \int_0^2 M_y \frac{\partial M_y}{\partial P}\,dy\right],$$

$$\frac{\partial M_x}{\partial P} = \frac{2x}{3}, \quad \frac{\partial M_y}{\partial P} = \frac{y}{3}.$$

For $P = 0$,
$$M_x = 15\,000x - \frac{10\,000x^2}{2} \quad \text{and}$$
$$M_y = 15\,000y - \frac{10\,000y^2}{2}.$$

$$\therefore \delta = \frac{1}{EI}\left[\int_0^1 \left(10\,000x^2 - \frac{10\,000x^3}{3}\right) dx \right.$$
$$\left. + \int_0^2 \left(5000y^2 - \frac{5000y^3}{3}\right) dy\right]$$
$$= \frac{10\,000}{EI}\left(\left[\frac{x^3}{3} - \frac{x^4}{12}\right]_0^1 + \left[\frac{y^3}{6} - \frac{y^4}{24}\right]_0^2\right)$$
$$= \frac{10\,000}{EI}\left(\frac{1}{3} - \frac{1}{12} + \frac{4}{3} - \frac{2}{3}\right)$$
$$= \frac{11}{12} \times \frac{10\,000}{EI}$$
$$= 2\cdot29 \text{ mm}.$$

8.4. The pin-jointed plane frame shown in Fig. 8.4 is pinned to a vertical wall at A and at B, and the angle $ACB = 90°$. Members AC and

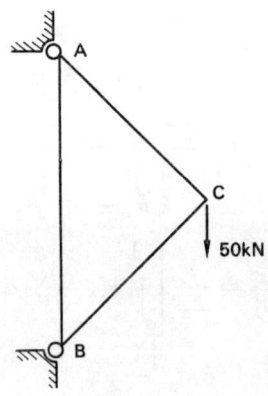

Fig. 8.4.

STRAIN ENERGY METHODS

BC are each 1 m long and 0·001 m² cross-sectional area. Find the vertical and horizontal deflections of C due to a vertical load of 50 kN acting vertically downwards at that point. ($E = 200$ GN/m².)

Solution

Vertical Deflection

Deflection $\delta = \partial U/\partial W$, where W is the load locating the point and direction of δ.

For a pin-jointed frame,

$$U = \sum \frac{P^2 L}{2AE}$$

where P is the tensile or compressive force in each bar.

$$\therefore \; \delta = \frac{\partial U}{\partial W} = \frac{\partial}{\partial W} \sum \frac{P^2 L}{2AE} = \sum \frac{PL}{AE} \frac{\partial P}{\partial W}.$$

In this expression P is the load carried by the member due to the actual external loading.

$\partial P/\partial W$ is the load carried by the member due to unit load acting at the point and in the direction for which the deflection is required.

TABLE 8.1. VERTICAL DEFLECTION

Member	Length (m)	P (kN)	$\dfrac{\partial P}{\partial W}$ (kN)	$\dfrac{P}{A}$	$\dfrac{PL}{A}\dfrac{\partial P}{\partial W}$
AC	1	$+\dfrac{50}{\sqrt{2}}$	$\dfrac{1}{\sqrt{2}}$	$\dfrac{50}{0\cdot001\sqrt{2}}$	25×10^6
BC	1	$-\dfrac{50}{\sqrt{2}}$	$\dfrac{-1}{\sqrt{2}}$	$\dfrac{-50}{0\cdot001\sqrt{2}}$	25×10^6

186   STRESS ANALYSIS PROBLEMS IN S.I. UNITS

$$\delta = \frac{1}{E}\sum \frac{PL}{A}\frac{\partial P}{\partial W}$$

$$= \frac{1}{200\times 10^9}\times 50\times 10^6 \text{ m}$$

$$= \frac{50}{200}\times \frac{10^9}{10^9} \text{ mm}$$

$$= \frac{1}{4} \text{ mm.}$$

Horizontal Deflection

Here it is necessary either to assume a horizontal force W which is put equal to zero or to use the statement in the previous section of this problem that $\partial P/\partial W$ represents the load in the member due to unit load in the direction of W and that P is the actual load in the member. Using the second method,

TABLE 8.2

Member	Length (m)	P (kN)	$\frac{\partial P}{\partial W}$ (kN)	$\frac{P}{A}$	$\frac{PL}{A}\frac{\partial P}{\partial W}$
AC	1	$+\frac{50}{\sqrt{2}}$	$\frac{1}{\sqrt{2}}$	$\frac{50}{\cdot 001\sqrt{2}}$	25×10^6
BC	1	$-\frac{50}{\sqrt{2}}$	$\frac{1}{\sqrt{2}}$	$\frac{-50}{\cdot 001\sqrt{2}}$	-25×10^6

$$\delta = \frac{1}{E}\sum \frac{PL}{A}\frac{\partial P}{\partial W}$$

$$= \frac{1}{E}[+25\times 10^6 - 25\times 10^6]$$

$$= 0.$$

STRAIN ENERGY METHODS 187

(*Note.* The method of tabulating the quantities shown is by far the easiest method of dealing with the calculations involved in applying Castigliano's first theorem to pin-jointed frames.)

8.5. The portal frame shown in Fig. 8.5 is pin-jointed to horizontal foundations at A and D. If there is a point load of 10 kN acting vertically down midway between B and C, find the horizontal component of the reaction at the foundation pin joints.

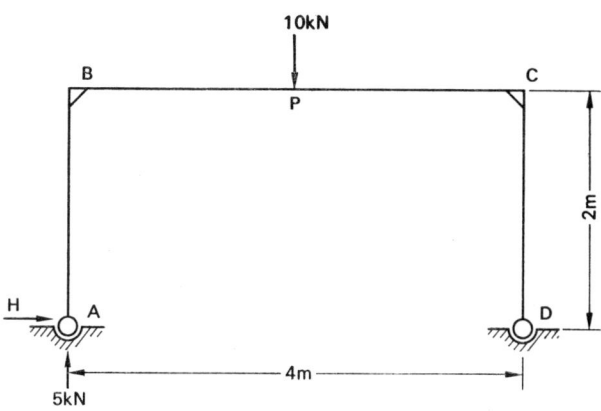

FIG. 8.5.

Solution

Referring to Fig. 8.5 and applying Castigliano's first theorem to find the horizontal deflection of point A,

$$\delta = \frac{\partial U}{\partial H},$$

but in fact point A is fixed so that $\delta = 0$ and an expression for $\partial U/\partial H$ can be written, equated to zero, and solved for H.

Since the frame is symmetrical, U can be evaluated for one half and then doubled.

Section AB:
$$M = -Hx$$
$$\frac{\partial M}{\partial H} = -x$$

for a section distance x vertically above A.

Section BP:
$$M = -2H + 5x$$
$$\frac{\partial M}{\partial H} = -2$$

for a section distance x from B in member BP.

Now, in general, $\quad \dfrac{\partial U}{\partial H} = \dfrac{1}{EI} \int M \dfrac{\partial M}{\partial H}\, dx.$

Carrying out the integration for each section, adding and doubling the result, gives

$$\frac{\partial U}{\partial H} = 2 \int_0^2 \frac{(-Hx)(-x)}{EI}\, dx + 2 \int_0^2 \frac{(-2H+5x)(-2)}{EI}\, dx$$

$$= \frac{2H}{3EI}\left[x^3\right]_0^2 + \frac{2}{EI}\left[4Hx - 5x^2\right]_0^2.$$

But $\quad \dfrac{\partial U}{\partial H} = 0.$

$$\therefore \; H = \tfrac{15}{8} = 1\cdot 875 \text{ kN}.$$

8.6. A horizontal I-section cantilever 150 mm deep and 130 mm wide has flanges and web all 15 mm thick. It is 1·2 m long and carries a vertical load of 30 kN at the free end. Find the shear strain energy in the whole beam and hence the deflection due to shearing only of the free end. ($G = 83$ GN/m² and $I = 2\times 10^{-5}$ m⁴.)

[London Univ.]

Solution

In general, shear strain energy $= \dfrac{q^2 \times \text{volume}}{2G}$.

The shear force is constant along this beam and the shear stress at a distance y from the neutral axis is

$$q_y = \frac{VA\bar{y}}{Ib}.$$

It is therefore necessary to evaluate the shear stress for an element of depth dy over which it can be taken as constant, finding the strain energy for this element over the whole length of beam, and then integrating over the full cross-section.

It is easiest in this problem, since the section is symmetrical, to carry out the calculation for (a) one flange, and (b) half the web and to double the sum.

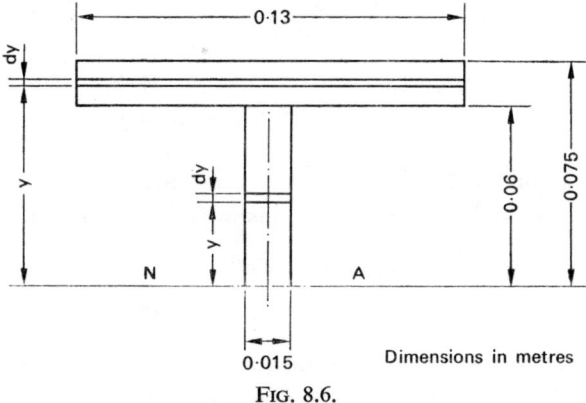

Fig. 8.6.

Refer to Fig. 8.6. For an element in the flange:

(a)
$$q_y = \frac{30 \times 10^3 \times (0{\cdot}075 - y)\,(0{\cdot}13) \times \left(y + \dfrac{(0{\cdot}075 - y)}{2}\right)}{2 \times 10^{-5} \times 0{\cdot}13}$$

$$= 7{\cdot}5 \times 10^8 \times (56 \times 10^{-4} - y^2).$$

Strain energy in element of flange

$$= \frac{[7\cdot 5\times 10^8\times (56\times 10^{-4}-y^2)]^2\times 1\cdot 2dy\times 0\cdot 13}{2\times 83\times 10^9}.$$

Integrating between limits $y = 0\cdot 06$ to $y = 0\cdot 075$ gives the strain energy in the whole flange.

$$= 0\cdot 01272 \text{ J}.$$

(b) For an element in the web,

$$q_y = 7\cdot 5\times 10^8(0\cdot 02116-y^2).$$

Strain energy in element

$$= \frac{[7\cdot 5\times 10^8(0\cdot 02116-y^2)]^2\times 0\cdot 015\times 1\cdot 2dy}{2\times 83\times 10^9}.$$

Integrating between limits, $y = 0$ to $y = 0\cdot 06$ gives
strain energy in half of web, $= 1\cdot 46$ J.
Total shear strain energy, $= 2\times (1\cdot 46+0\cdot 01272)$
$= 2\cdot 945$ J.

If the load at the free end sinks a distance δ the work done by the load $= \frac{1}{2}\times 30\times 10^3\delta$ and this work is stored as strain energy.

$$\therefore \tfrac{1}{2}\times 30\times 10^3\delta = 2\cdot 945,$$
$$\delta = 0\cdot 1963 \text{ mm}.$$

Note that this is the deflection due to shear. There will also be deflection due to bending (this component is calculated for the beam in this example in problem 8.17) and the total deflection will be the sum of the components. It frequently happens that one component is much larger than the other and the smaller can be neglected. In examination problems it is usual to ask for either the shear deflection or the bending deflection; this is done to reduce the problem to manageable proportions, and it does not mean that for the system concerned the other deflection component does not exist.

8.7. A load W falls through a height h on to the centre of a simply-supported beam of flexural rigidity EI and span L.

(a) What equivalent load W_e when steadily applied would produce the same deflection?

(b) What deflection would be produced by a mass of 600 kg falling 20 mm on to the mid-point of a span of 5·5 m if $EI = 187 \times 10^5$ Nm² and $g = 9·81$ m/s²?   [London Univ., 1957]

Solution

(a) Suppose the deflection caused by the impact $= \delta$.
Energy lost by the load $= W(h+\delta)$.
An equivalent load W_e slowly applied would do work $\tfrac{1}{2}W_e\delta$ in causing the same deflection.

$$\therefore \tfrac{1}{2}W_e\delta = W(h+\delta),$$

but the deflection under a central point load is known to be

$$\frac{W_e L^3}{48EI}.$$

Substituting this value for δ and solving the quadratic gives

$$W_e = W + \sqrt{\left(W^2 + \frac{96WhEI}{L^3}\right)}.$$

(b) This is merely a matter of substituting in the derived formula remembering that

$$W = 600 \times 9·81 = 5886 \text{ N}.$$

Hence $\delta = 7·784$ mm.

8.8. The pin-jointed plane frame in Fig. 8.7 has pin joints at A, B, C, D, and E only, members AD and EB cross with no joint. Vertical and horizontal members have cross-sectional areas of 10^{-3} m² and diagonal members have areas of 2×10^{-3} m².

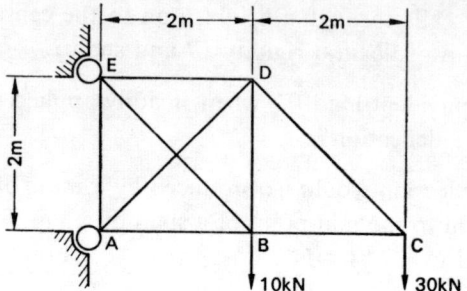

FIG. 8.7. Redundant (or statically indeterminate) frame.

Since the frame is redundant it is statically indeterminate; use Castigliano's second theorem to find the force in *BD* so that the rest of the frame can be solved by the methods of statics. ($E = 200$ GN/m².)

Solution

Referring to Fig. 8.7 one can assume that *BD* is the redundant member and that there is a tensile force R in it.

The second theorem states that $\partial U/\partial R$ represents the initial lack of fit of the member. If the initial fit was perfect $\partial U/\partial R = 0$ and this will provide the extra equation needed to solve the frame

$$\frac{\partial U}{\partial R} = \Sigma \frac{PL}{AE} \frac{\partial P}{\partial R} = 0,$$

where *P* is the force in a member due to the external loading plus the force due to the force *R* in member *BD*.

$$0 = \frac{1}{200 \times 10^9} [(6+4\sqrt{2})10^3 R + 40\sqrt{(2)} \times 10^6 + 40 \times 10^6],$$

$$R = -\frac{40(1+\sqrt{2})10^3}{6+4\sqrt{2}}$$

$$= -8 \cdot 28 \text{ kN},$$

i.e. a compressive force of 8·28 kN.

TABLE 8.3

Member	Length (m)	Area (m^2)	Load due to R (P_1)	Load due to externals (P_2)	$\partial P/\partial R$	$\dfrac{P}{A} = \dfrac{P_1+P_2}{A}$	$\dfrac{PL}{A}\dfrac{\partial P}{\partial R}$
BC	2	10^{-3}	0	-30×10^3	0	-30×10^6	0
CD	$2\sqrt{2}$	2×10^{-3}	0	$30\sqrt{2}\times 10^3$	0	$\dfrac{30}{\sqrt{2}}\times 10^6$	0
AB	2	10^{-3}	R	-40×10^3	1	$(R-40\times 10^3)10^3$	$(2R-80\times 10^3)10^3$
ED	2	10^{-3}	R	60×10^3	1	$(R+60\times 10^3)10^3$	$(2R+120\times 10^3)10^3$
AD	$2\sqrt{2}$	2×10^{-3}	$-\sqrt{2}R$	$-30\sqrt{2}\times 10^3$	$-\sqrt{2}$	$\dfrac{[-\sqrt{2}R-30\sqrt{2}\times 10^3]10^3}{2}$	$2\sqrt{2}\times 10^3(R+30\times 10^3)$
EB	$2\sqrt{2}$	2×10^{-3}	$-\sqrt{2}R$	$10\sqrt{2}\times 10^3$	$-\sqrt{2}$	$\dfrac{(-R+10^4)}{2}\sqrt{2}\times 10^3$	$(R-10^4)\,2\sqrt{2}\times 10^3$
BD	2	10^{-3}	R	0	1	$R\times 10^3$	$2R\times 10^3$

Problems

8.9. A hollow shaft 200 mm outside diameter and 130 mm inside diameter transmits 1350 kW at 16 rad/s. Calculate the stress at inner and outer surfaces and the strain energy per metre length of shaft. ($G = 83$ GN/m².)

[65·5 MN/m². 42·6 MN/m². 333 J.]

8.10. A hollow shaft subjected to a pure torque attains a maximum shearing stress q. If the strain energy per unit volume is $q^2/3G$, calculate the ratio of the shaft diameters.

Find the actual diameters for such a shaft required to transmit 3500 kW at 120 rev/min with a uniform torque when the energy stored is $20·7 \times 10^3$ J/m³. ($G = 80$ GN/m².) [London Univ.]

[$D/d = \sqrt{3}$. $D = 283$ mm. $d = 164$ mm.]

8.11. Given the strain energy in shearing stored in unit volume where there is a uniform shear stress q is $q^2/2G$, show from first principles that the strain energy in unit volume of a hollow shaft of diameters D and d, carrying a pure torque is

$$\frac{q_{max}^2}{4G}\left[\frac{D^2+d^2}{D^2}\right].$$

Hint: Consider the strain energy in a hollow cylindrical element of shaft of thickness dr, unit length and radius r over the thickness of which the shear stress q will be constant and related to the maximum shear stress q_M by $q = rq_M/(D/2)$.

From the given expression the strain energy in this element can be written and integration carried out between limits $d/2$ to $D/2$ to obtain the strain energy in unit length. It is then only necessary to divide by the volume of unit length.

8.12. A uniform cantilever of length 1·2 m, $I_{NA} = 2 \times 10^{-5}$ m⁴, has a pure torque of 10^4 N m applied at the free end about a horizontal axis perpendicular to the length of the cantilever. Find the slope and deflection at the free end. ($E = 200$ GN/m².)

[Slope $= 3 \times 10^{-3}$. Deflection $= 1\cdot 8$ mm.]

Notes

(1) Since the slope is small it will be approximately equal to the angular displacement θ in radians.

(2) The angular displacement at the point of application of a torque T_v is

$$\frac{\partial U}{\partial T} = \frac{\partial}{\partial T} \int_0^L \frac{M^2 \, dx}{2EI} = \frac{1}{EI} \int_0^L M \frac{\partial M}{\partial T} \, dx,$$

where M is the bending moment at a general point located at x along the beam.

(3) To find the displacement of the free end, since there is no force there, an imaginary force $P(=0)$ must be applied and the displacement found from

$$\delta = \frac{\partial U}{\partial P} = \frac{1}{EI} \int_0^L M \frac{\partial M}{\partial P} \, dx.$$

8.13. Find the slope at the free end of a horizontal cantilever $1\cdot 2$ m long, second moment 2×10^{-5} m⁴ due to a vertical load of 30 kN at the end. ($E = 203$ GN/m².)

[$5\cdot 32 \times 10^{-3}$ rad.]

8.14. Find the ratio of the bending strain energy U_D in a simply-supported beam carrying a uniformly distributed load to the bending strain energy U_C in the same beam carrying a central point load if the loads are such that the maximum bending stresses are equal.

[$U_D/U_C = 1\cdot 6$.]

8.15. A horizontal cantilever of length L, area A, second moment I, carries a point load P at the free end. Use Castigliano's first theorem to find the deflection of the free end of the beam.

$$\left[\delta = \frac{PL^3}{3EI}.\right]$$

8.16. Calculate, using Castigliano's theorem, (a) the deflection, and (b) the slope, at the free end of the horizontal cantilever described in problem 8.15, if, instead of the point load P, it carries a uniformly distributed load w per unit length applied over the whole length.

$$\left[\delta = \frac{wL^4}{8EI}. \quad \theta = \frac{wL^3}{6EI}.\right]$$

8.17. Find the total strain energy due to bending in the I-section cantilever described in example 8.6, if $E = 203$ GN/m². Hence find the deflection due to bending of the free end and the ratio of the bending to the shear deflection.

[63·9 J. $\delta = 4\cdot26$ mm. 21·7 : 1.]

8.18. A bar of uniform flexural rigidity EI and total effective length $6a$ is bent to the shape shown in Fig. 8.8 and fixed to a solid wall at D so that AB and CD are horizontal and CB is vertical. A load W is

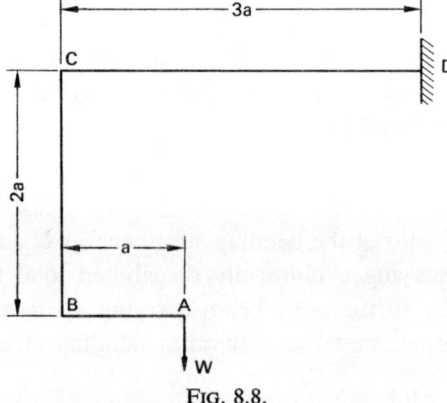

Fig. 8.8.

fixed at *A*. Derive expressions for the vertical deflections, due to bending only, of points *A* and *C* in terms of *W*, *a*, and *EI*.

[London Univ.]

$$\left[\delta_C = \frac{3Wa^2}{EI}. \quad \delta_A = \frac{16Wa^3}{3EI}.\right]$$

(*Note*. In fact there would be vertical deflection components due to shear in *CD* and *AB* and to tension in *BC*, but these are likely to be small compared with the bending deflections.)

8.19. Find the horizontal and vertical deflections of the free end of a circular cross-section steel bar bent to form a quarter circle of radius

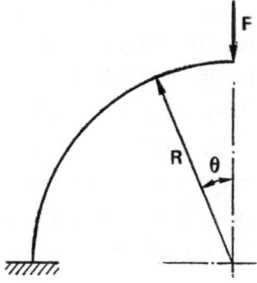

Fig. 8.9.

R and fastened to a rigid horizontal base (Fig. 8.9) at one end, when a vertical force *F* is applied at the free end.

$$\left[\delta_v = \frac{\pi F R^3}{4EI}. \quad \delta_n = \frac{FR^3}{2EI}.\right]$$

$\Bigg($ *Note.* The general expression for strain energy in bending, $\int_0^L \dfrac{M^2\,dx}{2EI}$, involves the bending moment *M* at an element of length *dx* located at *x* along the beam. In this problem $dx = R\,d\theta$, and the general expression for *M* in terms of *F*, *R*, and θ is $M = FR\sin\theta$.

$$\therefore\ U = \int_0^{\pi/2} \frac{(FR\sin\theta)^2}{2EI} R\,d\theta \quad \text{and} \quad \delta = \frac{\partial U}{\partial F}.\Bigg)$$

8.20. A semicircular arch of radius R is pinned to two rigid supports A and B at the same level. Find the magnitude of the horizontal reactions H at the supports caused by a point load W at the centre of the arch.

[W/π.]

8.21. Find the deflection, caused by the 10 kN force, of the point P in the portal frame of example 8.5 if the frame is made of rectangular cross-section members 50 mm wide and 75 mm deep and $E = 200$ GN/m².

[16·55 mm.]

8.22. A solid bar of circular cross-section, diameter d, is fastened rigidly at the base and projects vertically upwards for a distance L. The remaining length is bent into a quadrant of mean radius R and a load W is hung on the free end. The dimensions R and L are large compared to d. Derive expressions for the horizontal and vertical deflections due to bending of the free end and calculate the value of vertical deflection if $d = 50$ mm, $L = 1\cdot 2$ m, $R = 0\cdot 9$ m, $W = 450$ N, $E = 207$ GN/m².

$$\left[\delta_V = \frac{WR^2}{EI}\left(\frac{\pi R}{4}+L\right). \quad \delta_H = \frac{WR}{2EI}(R^2+L^2). \quad \delta_V = 10\cdot 95 \text{ mm.} \right]$$

8.23. A beam 150 mm wide, 250 mm deep, is simply supported at the ends over a span of 3 m. A load weighing 1800 N falls a distance of 90 mm on to the middle of the beam. Neglecting loss of energy at impact, find the maximum instantaneous stress produced in the beam given that $I = 8\cdot 75 \times 10^{-5}$ m⁴ and $E = 200$ GN/m².

[London Univ.]

[$f = 109\cdot 5$ MN/m².]

8.24. What is the maximum deflection of the beam under the conditions of problem 8.23?

[3·29 mm.]

8.25. Figure 8.10 shows a pin-jointed crane; the wall is vertical and all members except AB are 2 m long and the cross-sectional areas are $AB = 8 \times 10^{-4}$ m², $BC = 10 \times 10^{-4}$ m², $CD = 13 \times 10^{-4}$ m², $CA =$

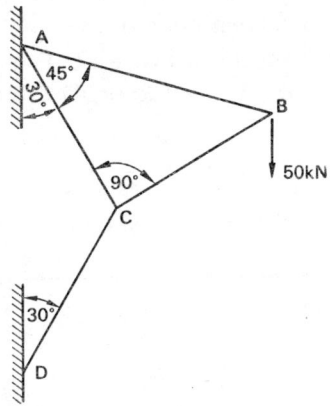

Fig. 8.10.

6×10^{-4} m². Find the vertical deflection of joint B due to the 50 kN load. ($E = 200$ GN/m².) [London Univ., 1965]

[3·74 mm.]

8.26. Figure 8.11 shows a pin-jointed plane frame in which all tension members carry stresses of 90 MN/m² and all compression members

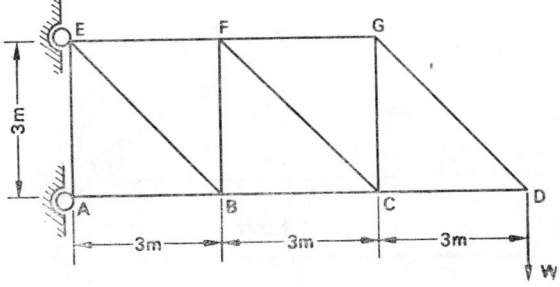

Fig. 8.11.

carry 75 MN/m² due to the vertical load W acting at point D. If E for the frame members is 200 GN/m², find the vertical deflection of D.

[21·15 mm.]

8.27. Calculate the vertical deflection of the 3 kN load applied to the plane pin-jointed frame shown in Fig. 8.12. The cross-sectional area of all members is 1.9×10^{-3} m² and $E = 207$ GN/m².

[London Univ.]

[0·227 mm.]

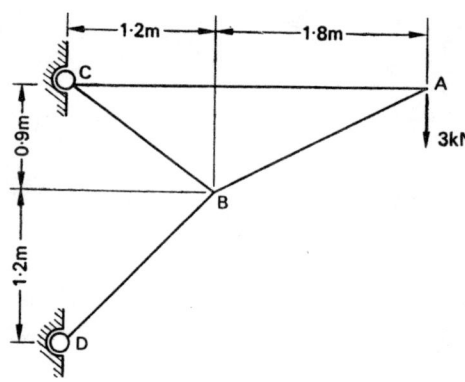

Fig. 8.12.

8.28. The pin-jointed truss shown in Fig. 8.13, has a span of 4 m. A load of 9 kN acts vertically down at the pin joint A. Find the ver-

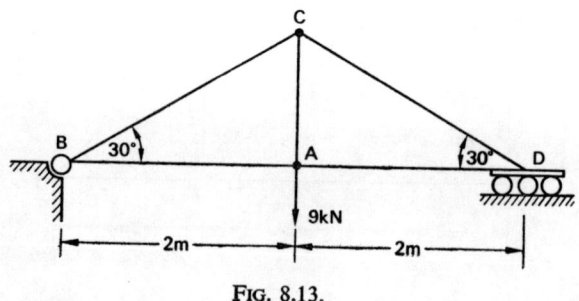

Fig. 8.13.

tical deflection of this point due to the load. Cross-sectional area of all members 2×10^{-3} m² and $E = 200$ GN/m².

[0·16 mm.]

8.29. A square pin-jointed plane frame *ABCD* of side 2 m is braced by two diagonal members which are not joined where they cross. A tensile force of 10 kN is applied to joints *A* and *C* along the di-

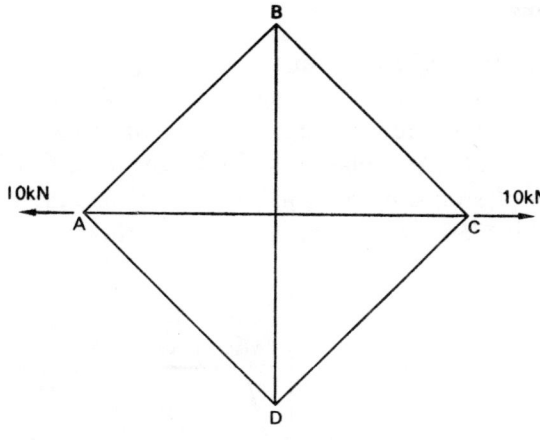

FIG. 8.14.

agonal as shown in Fig. 8.14. Using Castigliano's second theorem, find the force in member *AC*. Area of each member 2×10^{-3} m² and $E = 200$ GN/m².

[7·07 kN tensile.]

8.30. A cantilever protrudes horizontally from a solid wall. The overhang is 500 mm and a point load of 100 kN acts vertically down at the free end. If the section of the beam is a square of 100 mm side and if $G = 80$ GN/m², find the deflection due to shear of the free end.

[0·075 mm.]

CHAPTER 9

Elementary Plastic Stress Analysis

Definitions and Theory

(a) Most ductile engineering metals have a stress/strain curve with a linear portion up to some limiting value and, for larger stresses, a non-linear portion where very large increases of strain take place for a small increase of stress. Plastic theory makes the assumption that, in the region of strains permissible in real systems, the stress/strain curve for a ductile metal can be idealized to the form of Fig. 9.1.

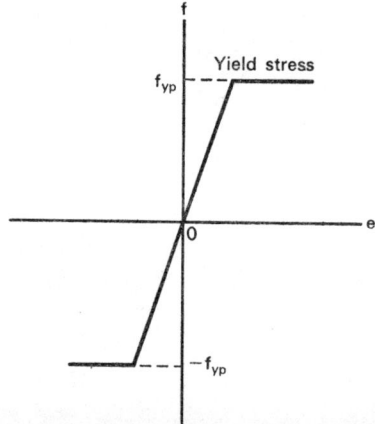

FIG. 9.1. Idealized stress/strain graph for a ductile material.

The assumption is therefore made that loading on a member which produced a stress slightly above the yield stress f_{yp} will cause the member to continue to strain until complete failure by gross deformation takes place or until small deformations of the system cause a redistribution of the loading and the stress in the member concerned falls back below the yield point.

(b) In the method of "limit design", the proportions of members are so chosen that under the loading system expected to be applied each member would be stressed to the same proportion of the yield stress. This may be done by assuming that the yield-point stress is acting in every member—as it would do in the limiting case—and then calculating the maximum external loading which could be applied. In order that gross deformation and collapse does not in fact take place, the actual load permitted will be less by some "factor of safety" or "load factor" than the limit load.

(c) The torsion test on circular shafts provides a simple case of elastic/plastic deformation which can be easily analysed. In Chapter 5 it was shown that for elastic torsion of a circular shaft the relation

$$\frac{T}{J} = \frac{q}{r} = \frac{G\theta}{L}$$

existed between torque T, polar second moment J, shear stress q at radius r, rigidity modulus G, angle of twist in radians θ, and length L.

If T is sufficiently great, q will reach q_{yp}, the yield value of shear stress, and it will do so first at the outer surface of the shaft,

i.e. $$q_{yp} = \frac{DT}{2J}$$

where D is the outside diameter.

If a greater torque is applied, plastic deformation will take place at the surface though the inner parts of the shaft will still be stressed below the elastic limit. If the shaft is still elastic up to a radius r_1, and plastic thereafter, the total torque carried T_t will be divided between plastic and elastic portions,

i.e. $$T_t = T_{\text{elastic}} + T_{\text{plastic}},$$

where

$$T_{\text{elastic}} = \frac{q_{yp}J}{r_1},$$

$$T_{\text{plastic}} = 2\pi q_{yp} \int_{r_1}^{D/2} r^2 \, dr.$$

In evaluating the twist of the shaft the assumption is made that a line, radial before twisting, remains radial in spite of elastic/plastic twisting.

Thus the angle of twist θ calculated for the elastic "core" from

$$\theta = \frac{q_{yp}L}{r_1 G}$$

will apply to the shaft as a whole.

(d) The concept of plastic bending of beams assumes:

(1) That the ideal stress–strain curve (Fig. 9.1) is geometrically similar on both sides of the axes, i.e. f_{yp} in compression $= f_{yp}$ in tension.
(2) Cross-sections, plane before plastic bending, remain plane after bending.
(3) Shear stresses in the beam have negligible effect on plastic yield.

Figure 9.2a shows the distribution of bending stress in accordance with the theory of bending dealt with in Chapter 4 where the elastic limit is not exceeded; Fig. 9.2b shows a typical distribution when the applied moment M is large enough to cause the yield stress to be developed some way into the beam on both tension and compression sides of the neutral layer; Fig. 9.2c shows the extreme case where plastic deformation is taking place at all layers of the cross-section and the stress is equal to f_{yp} throughout both the tensile and compressive sides of the neutral axis. This corresponds to the maximum moment which the beam can withstand and the point along the beam where it occurs is called a "plastic hinge".

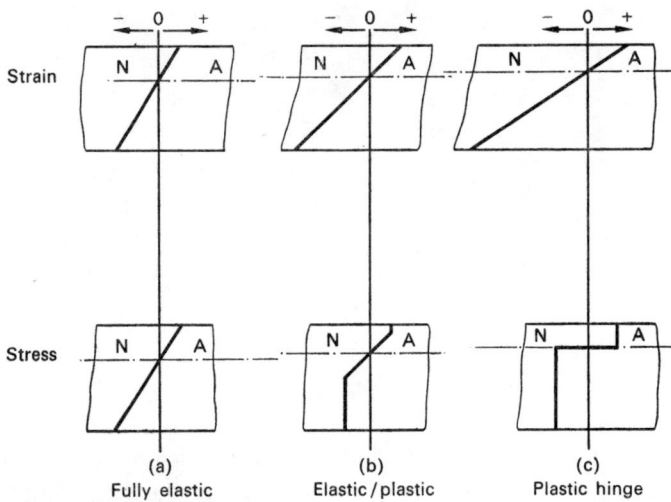

Fig. 9.2. Distribution of stress and strain in a beam of cross-section not symmetrical about a horizontal axis.

Under plastic hinge conditions:

(1) The neutral axis divides the cross-section, total area A, into compressive and tensile sections A_1 and A_2, such that $A_1 = A_2 = \tfrac{1}{2}A$.

(2) The maximum bending moment M_p is

$$M_p = f_{yp} \times \tfrac{1}{2}A(\bar{y}_1 + \bar{y}_2),$$

where $\bar{y}_1$ and $\bar{y}_2$ are the distances from the neutral axis to the centroids of areas A_1 and A_2.

Worked Examples

9.1. Three steel tie-rods, each of cross-sectional area 1000 mm², are arranged to form a plane pin-jointed redundant frame as shown in Fig. 9.3. A load of P newtons vertically downwards is applied at joint D.

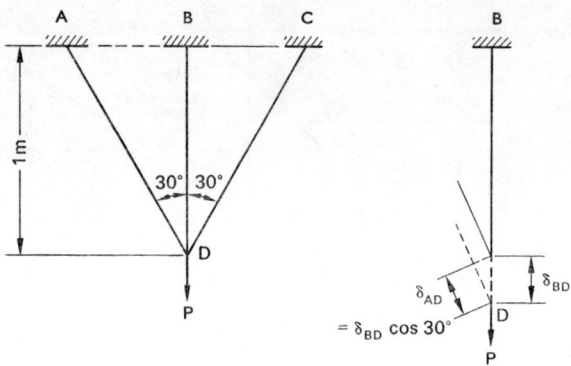

FIG. 9.3.

Find (a) the least value of P that will cause yielding in some of the members, (b) the stress in each member under this load, and (c) the least load that will cause all members to yield and the frame to collapse. ($f_{yp} = 240$ MN/m² and $E = 200$ GN/m².)

Solution

(a) Consider the equilibrium of point D, Fig. 9.3. It is clear that the force in AD, F_{AD}, must equal the force in CD, F_{CD}. Also resolving forces vertically, upwards positive,

$$F_{BD} + F_{AD} \cos 30° + F_{CD} \cos 30° - P = 0.$$
$$\therefore \quad F_{BD} + 2F_{AD} \cos 30° = P.$$

There are still two unknowns, and the system is statically indeterminate. However, considering deformations for compatibility if BD extends δ_{BD} and AD extends δ_{AD},

$$\delta_{AD} = \delta_{BD} \cos 30°.$$

Now
$$E = \frac{f}{e}$$
$$= \frac{f}{\delta/L}.$$

$$\therefore\ \delta = \frac{fL}{E}$$

$$= \frac{FL}{AE}.$$

$$\therefore\ \delta_{AD} = \frac{F_{AD}\frac{2}{\sqrt{3}}}{10^{-3} \times 2 \times 10^{11}},$$

$$\delta_{BD} = \frac{F_{BD} \times 1}{10^{-3} \times 2 \times 10^{11}}.$$

Substituting gives

$$F_{AD} = \tfrac{3}{4} F_{BD} \tag{1}$$

$$\therefore\ F_{BD} + \tfrac{3}{2} F_{BD} \cos 30° = P, \quad \text{giving} \quad P = 2 \cdot 299 F_{BD}. \tag{2}$$

From eqn. (1) is is clear that BD carries the greatest force and since all rods have the same area it will reach yield conditions first.

When stress in $BD = 240 \times 10^6$ N/m², $P = 550$ kN and this is the least value that will cause yielding in any member.

Under this load, stress in AD and $CD = \tfrac{3}{4} \times 240 \times 10^6$ N/m²
$= 180$ MN/m².

The above solutions do not require the application of the concept of a plastic limit, but part (c) can be very simply solved in this way by assuming that P is increased until BD yields and goes on yielding at constant stress until the stress in AD and CD also reaches the yield when complete collapse occurs.

(c) If stress in every rod is 240 MN/m², force in every member is $240 \times 10^6 \times 10^{-3}$ N $= 240$ kN.

But $\quad P = F_{BD} + 2F_{AD} \cos 30°$ still applies.

$$\therefore\ P = 240 \times 10^3 + 2 \times 240 \times 10^3 \times \frac{\sqrt{3}}{2}$$

$$= (1 + \sqrt{3}) \times 240 \text{ kN}.$$

$$P = 655 \text{ kN}.$$

Note that in part (c) it was necessary for all members to yield before the system collapsed. In general, however, the system will collapse if enough members yield to render the framework non-rigid.

9.2. A rigid horizontal beam 2 m long is supported by three vertical steel tie-rods each 500 mm long and 6×10^{-4} m² cross-sectional area. If the bars are pinned to the beam, one at each end and one at the mid-point, what vertical load could be applied to the beam at a point 0·4 m inside one of the end tie-rods without causing the system to come within a safety factor of 2 of collapse by yielding of the tie-rods? ($f_{yp} = 450$ MN/m².)

Solution

Refer to Fig. 9.4.

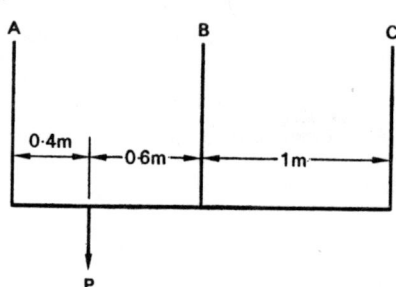

Fig. 9.4.

If rod A yields, tension in B and compression in C could still maintain equilibrium, but if B or C also yields, the system will collapse. Suppose B also yields:

$$F_A = F_B = 450 \times 6 \times 10^{-4} \times 10^6 \text{ N} = 270 \text{ kN}.$$

Considering equilibrium of the beam,

$$F_A + F_B + F_C = P \quad \text{(assuming all rods carry tension)}.$$

And taking moments about C, clockwise positive,

$$2F_A + F_B - 1 \cdot 6P = 0.$$

Substituting values F_A and F_B,

$$2\times 270\times 10^3 + 270\times 10^3 = 1\cdot 6P,$$
$$P = 506\cdot 25 \text{ kN}.$$

Hence $\qquad\qquad 270 + 270 + F_C = 506\cdot 25$

and $\qquad\qquad\qquad\qquad F_C = -33\cdot 75 \text{ kN},$

i.e. it is in compression and well below collapse by yield (though in practice it might fail by buckling).

Thus a load of 506·25 kN will cause collapse, and if a factor of safety of 2 is to be applied, P must not exceed 253·125 kN.

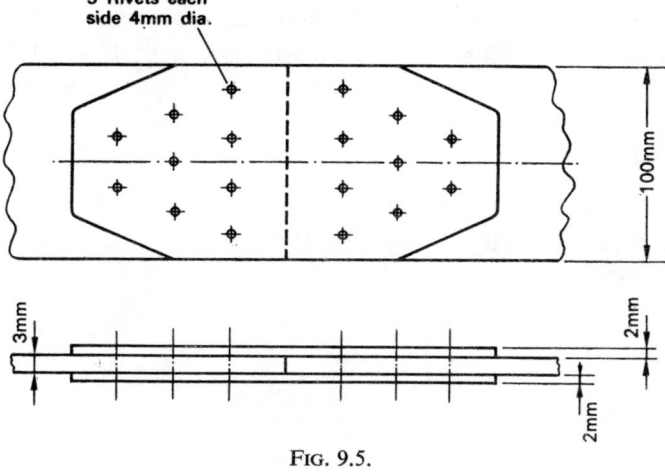

FIG. 9.5.

9.3. The riveted joint between steel plates shown in Fig. 9.5 has mild steel rivets 4 mm diameter. Calculate the maximum safe load using a factor of safety of 3 against failure by yielding of all the rivets if the yield stress of the rivets in shear is 138 MN/m².

Solution

The rivets are in double shear and if the joint is to fail, all must yield along both shear planes.

Total area sheared $= 2 \times 9 \times \pi \times \dfrac{(4 \times 10^{-3})^2}{4}$ m².

Total force $= 138 \times 10^6 \times 2 \times 9 \times \pi \times \dfrac{16 \times 10^{-6}}{4}$ N

$= 31 \cdot 23$ kN.

For a factor of safety of 3, maximum safe load $= 10 \cdot 41$ kN.

9.4. A case-hardened steel shaft 25 mm diameter has a case depth of 1·5 mm. Assuming the case remains perfectly elastic up to its failing stress in shear of 310 MN/m² and that the core becomes perfectly plastic at a shearing stress of 185 MN/m², calculate (a) the torque to cause a failure of the case in torsion, and (b) the angle of twist per metre length at failure. Both case and core have $G = 80$ GN/m².
[London Univ.]

Solution

One can assume that the case will carry torque as if it were a hollow shaft. When $q = 310$ MN/m² at outside of case, the torque carried by the case,

$$T_c = \dfrac{qJ}{D/2}$$

$$= \dfrac{310 \times 10^6 J}{125 \times 10^{-4}},$$

$$J = \dfrac{\pi}{32}(25^4 - 22^4)\,10^{-12}.$$

$$\therefore\ T_c = 381 \cdot 5 \text{ Nm}.$$

On the assumption that radial lines remain radial,

$$\dfrac{q}{r} = \dfrac{G\theta}{L}$$

will apply both to the case and to the elastic region of the core even though the outer layers of the core have become plastic. If the core is plastic beyond radius r_1, where the stress $q_1 = 185$ MN/m², and

$$\frac{q_1}{r_1} = \frac{q}{D/2},$$

where q is the stress at the outside of the core because both are equal to $G\theta/L$,

then, $\quad \dfrac{q_1}{r_1} = \dfrac{310 \times 10^6}{0.0125} \quad$ and $\quad r_1 = \dfrac{185 \times 10^6 \times 0.0125}{310 \times 10^6}$ m

$$= 7.46 \text{ mm}.$$

Torque carried by elastic region of core $= T_E$,

$$T_E = \frac{q_1 J_1}{r_1}$$

where J_1 is the second moment of the elastic portion,

$$= \frac{185 \times 10^6}{7.46 \times 10^{-3}} \times \frac{\pi}{2} (7.46 \times 10^{-3})^4$$

$$= 121.8 \text{ N m}.$$

Torque carried by plastic region of core $= T_P$.
Throughout this region, stress $= q_{yp}$.
On an elemental ring at radius r_1, radial thickness dr, the shear force is $q_{yp} 2\pi r \, dr$.
Moment of this force about axis $= 2\pi q_{yp} r^2 \, dr$.
Therefore total torque carried by this part of core

$$T_P = \int_{7.46 \times 10^{-3}}^{11 \times 10^{-3}} (2\pi \times 185 \times 10^6 r^2 \, dr) = 356 \text{ N m}.$$

Therefore total torque $= T_C + T_E + T_P = 859.3$ N m.
Twist can be calculated either from the case or the elastic core.
Using the case

$$\theta = \frac{310 \times 10^6 \times 1 \times 57.3}{80 \times 10^9 \times 12.5 \times 10^{-3}},$$

$$\theta = 17.75°.$$

9.5. A steel beam of rectangular cross-section 76 mm deep, 32 mm wide, is 1·2 m long and simply supported at the ends. A point load W at mid-span is just large enough to cause yielding to begin at the outer layers of the beam. If the yield stress of the material is 280 MN/m² find the magnitude of W.

What load would be necessary to cause the yielding to spread to a depth of 13 mm from the surface at mid-span and how far along the surface of the beam will the yielding spread?

[London Univ.]

Solution

The position of the neutral axis for elastic bending is through the centroid of the section; for plastic hinge conditions it will be such that it divides the section into equal areas; for elastic/plastic bending it will be between these two positions and located such that the total tensile force (on one side) is exactly equal to the total compressive force (on the other side).

In this problem, since the section is symmetrical about the horizontal centre line, the neutral axis will remain in the same position under all three conditions.

When yielding begins at the surface,

$$M = \frac{fI}{y}$$

$$= \frac{280 \times 10^6 \times 3 \cdot 2 \times 7 \cdot 6^3 \times 10^{-8}}{0 \cdot 038 \times 12} = 8 \cdot 64 \text{ kN m.}$$

But for a simply-supported beam with central point load,

$$M = \frac{WL}{4}.$$

$$\therefore W = \frac{4M}{L}$$

$$= 28 \cdot 8 \text{ kN.}$$

ELEMENTARY PLASTIC STRESS ANALYSIS

After yielding to a depth of 13 mm the moment carried by the core which is still elastic can be obtained by the simple bending equation as before,

$$M_c = \frac{280 \times 10^6 \times I}{0.025},$$

$$I = \frac{0.032 \times 0.05^3}{12}$$

since the elastic portion is 50 mm deep.

$$\therefore M_c = 3.75 \text{ kN m.}$$

In the plastic region, stress is assumed constant at 280 MN/m². Force in each plastic region of the cross-section

$$= 280 \times 10^6 \times 0.013 \times 0.032 \text{ N.}$$

Since the compressive and tensile forces are equal and opposite, they form a couple of moment arm 63 mm.

$$M_P = \text{force} \times 0.063$$
$$= 7.34 \text{ kN m.}$$

Total moment $= 7.34 + 3.75 = 11.09$ kN m.
This must be equal to the applied moment $WL/4$.

$$W = \frac{4}{1.2} \times 11.09 \times 10^3$$
$$= 37 \text{ kN.}$$

The first expression in the solution gives the moment necessary to initiate yielding at the surface; therefore wherever the bending moment is equal to or greater than 8·64 kN m, yielding will take place. Suppose yield commences at x from a support,

$$\frac{37 \times 10^3 x}{2} = 8.64 \times 10^3,$$

$$x = \frac{8.64 \times 2}{37}$$

$$= 0.466 \text{ m,}$$

i.e. yielding occurs at the surface for a distance of

$$1\cdot 2 - 2 \times 0\cdot 466 \text{ m}$$

$$= 268 \text{ mm}.$$

9.6. A steel T-section beam is used with the web vertical and the flange uppermost. Both flange and web are 19 mm thick, while the overall depth is 200 mm and the width 125 mm. A bending moment is applied which causes yielding in the bottom 50 mm of the web. If the yield stress is 278 MN/m² find (a) the position of the neutral axis, (b) the stress at the top of the flange, (c) the moment of resistance of the section. [London Univ., B.Sc. II]

Solution

Referring to Fig. 9.6, suppose the neutral axis is at a distance y mm

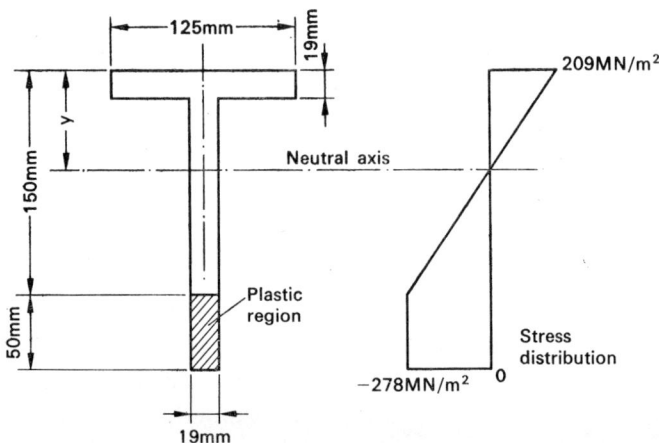

Fig. 9.6.

from the top of the flange. In example 9.5 it was stated that for elastic/plastic bending the neutral axis will be in a position which makes the total force on the cross-section equal to zero.

ELEMENTARY PLASTIC STRESS ANALYSIS

Stress at top of flange

$$= 278 \times 10^6 \times \left(\frac{y}{150-y}\right).$$

Stress at bottom of flange

$$= 278 \times 10^6 \times \left(\frac{y-19}{150-y}\right).$$

Average stress in flange

$$= \frac{1}{2} \times 278 \times 10^6 \times \left(\frac{2y-19}{150-y}\right).$$

Force in flange

$$= 139 \times 10^6 \times \left(\frac{2y-19}{150-y}\right) \times 19 \times 10^{-3} \times 125 \times 10^{-3}.$$

Average stress in section of web above neutral axis

$$= \frac{1}{2} \times 278 \times 10^6 \times \left(\frac{y-19}{150-y}\right).$$

Force in section of web above neutral axis

$$= 139 \times 10^6 \times \left(\frac{y-19}{150-y}\right) \times (y-19) \times 10^{-3} \times 19 \times 10^{-3}.$$

Stress in plastic area of web

$$= 278 \times 10^6 \text{ N/m}^2.$$

Force in plastic area of web

$$= 278 \times 10^6 \times 50 \times 10^{-3} \times 19 \times 10^{-3} \text{ N}.$$

Force in elastic area of web below neutral axis

$$= \tfrac{1}{2} \times 278 \times 10^6 \times 19 \times 10^{-3} \times (150-y) \times 10^{-3}.$$

Forces above and below neutral axis are equal:

$$\frac{139\times 19}{150-y}[(2y-19)\times 125+(y-19)^2] = 278\times 19\left[50+\frac{150-y}{2}\right],$$

$$250y-2375+y^2-38y+361 = (250-y)(150-y),$$
$$y^2+212y-2014 = 37\,500-400y+y^2$$
$$612y = 39\,514,$$
$$y = \frac{39\,514}{612}$$
$$= 64\cdot 6 \text{ mm}.$$

Stress at top of flange $= 278\times 10^6 \times \dfrac{64\cdot 6}{150-64\cdot 6}$

$$= 278\times \frac{64\cdot 6}{85\cdot 4}\times 10^6 \text{ N/m}^2$$

$$= 209 \text{ MN/m}^2.$$

Total resisting moment will be the moment about the neutral axis of the forces calculated above. The resultant force on the web will act at a point slightly above mid-depth, but a simplification with negligible error will be obtained by assuming it is exactly at the mid-point.

Moment of force in flange

$$= 139\times 19\times 125\times \frac{110\cdot 2}{85\cdot 4}\times 55\cdot 1\times 10^{-3} \text{ N m}.$$

$$= 23\cdot 5 \text{ kN m}.$$

Moment of force above neutral axis in elastic web

$$= 139\times 19\times \frac{45\cdot 6}{85\cdot 4}\times 45\cdot 6\times \tfrac{2}{3}\times 45\cdot 6\times 10^{-3}$$

$$= 1\cdot 95 \text{ kN m}.$$

Moment of force below neutral axis in elastic web

$$= 139\times 19\times 85\cdot 4\times \tfrac{2}{3}\times 85\cdot 4\times 10^{-3}$$
$$= 12\cdot 9 \text{ kN m}.$$

ELEMENTARY PLASTIC STRESS ANALYSIS

Moment of force below neutral axis in plastic web

$$= 278 \times 50 \times 19 \times (175 - 64 \cdot 6) \times 10^{-3}$$
$$= 29 \cdot 0 \text{ kN m.}$$

Total resisting moment

$$= 23 \cdot 5 + 1 \cdot 95 + 12 \cdot 9 + 29 \cdot 0$$
$$= 67 \cdot 35 \text{ kN m.}$$

9.7. For calculating the resisting moment M which a beam can develop, the simple theory of elastic bending uses the expression $M = fI/y_{max}$ or, since I/y_{max} is a quantity dependent only on the beam cross-sectional geometry, I/y_{max} is sometimes referred to as the section modulus Z and tabulated as a single value.

By considering a square cross-section of 0·1 m side, confirm the statement that under fully plastic conditions the plastic moment of resistance M_p of a cross-section symmetrical about its horizontal axis is given by $M_p = f_y S$, where S is the first moment of area of the cross-section about the axis. For this section find the shape factor $v = S/Z$.

Solution

$$I = \frac{bd^3}{12} \quad \text{and} \quad y = \frac{d}{2}.$$

$$\therefore Z = \frac{I}{y} = \frac{bd^2}{6}.$$

Under plastic conditions the stress over the whole section above the horizontal axis is f_y.

Therefore force on this section $= f_y \dfrac{bd}{2}$.

Moment of force about axis $= f_y \dfrac{bd}{2} \times \dfrac{d}{4}$.

Moment of force on area below axis is also $f_y \dfrac{bd}{2} \times \dfrac{d}{4}$.

$$\therefore M_p = 2f_y \dfrac{bd}{2} \times \dfrac{d}{4}$$

$$= f_y \dfrac{bd^2}{4}.$$

First moment of area of a horizontal element dy high at y from axis $= by\, dy$.

Therefore first moment of whole area $S = \displaystyle\int_{-d/2}^{+d/2} by\, dy = \left[\dfrac{by^2}{2}\right]_{-d/2}^{d/2}$

$$S = \dfrac{b}{2}\left(\dfrac{d^2}{4} + \dfrac{d^2}{4}\right),$$

$$S = \dfrac{bd^2}{4}.$$

$$\therefore M_p = f_y S.$$

$$v = \dfrac{S}{Z} = \dfrac{bd^2}{4} \times \dfrac{6}{bd^2} = 1\cdot 5.$$

9.8. Illustrate with sketches the plastic collapse conditions for a simply-supported beam with uniformly distributed loading and for a built-in beam. Calculate the uniformly distributed loading necessary to cause collapse of the built-in beam if the plastic moment M_p is 85·4 kN m and the span is 5 m.

Solution

Figure 9.7 shows how the formation of a plastic hinge at the midspan of a simply-supported beam will allow rotation of each half and complete collapse. For the built-in beam, however, it will be necessary for plastic hinges to form at both ends and in the middle before collapse can occur.

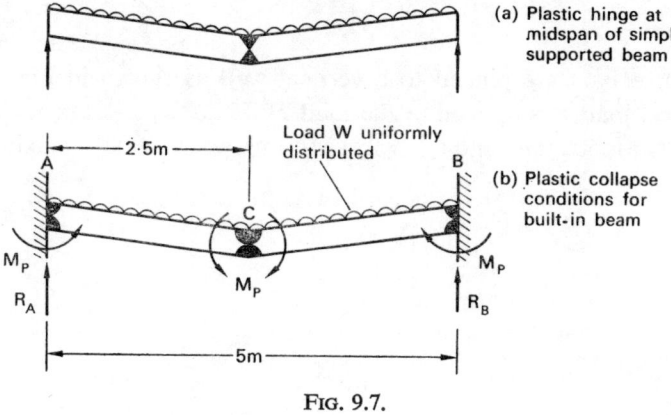

Fig. 9.7.

Considering the section AC of the built-in beam and taking moments about C, clockwise positive,

$$0 = -2M_p + R_A \times 2 \cdot 5 - \frac{W}{2} \times \frac{2 \cdot 5}{2}.$$

Considering forces; upwards positive,

$$0 = R_A - \frac{W}{2}$$

since there is no shear force across the hinge.

$$\therefore \ 0 = -2M_p + \frac{W}{2} \times 2 \cdot 5 - \frac{W}{2} \times \frac{2 \cdot 5}{2},$$

$$-2M_p = -\frac{2 \cdot 5 W}{4},$$

$$W = \frac{8}{2 \cdot 5} M_p$$

$$= \frac{8}{2 \cdot 5} \times 85 \cdot 4 \times 10^3 \text{ N}$$

$$= 273 \text{ kN}.$$

Problems

9.9. Three bars are pinned to a vertical wall as shown in Fig. 9.8. A vertical load P is applied at the joint D. If the cross-sectional area of each bar is 600 mm² and $f_{yp} = 275$ MN/m², what is (a) the maximum

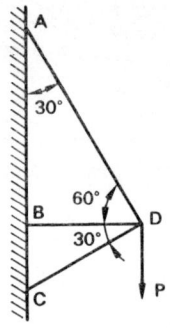

Fig. 9.8.

value of P if the structure is to have a factor of safety of 2 against collapse by yielding, and (b) the stress in each member at this value of P?

[(a) 113 kN. (b) Stress in AD and $CD = 275$ MN/m². Stress in $BD = 100$ MN/m².]

9.10. What would be the maximum safe value for P, using a load factor of 2, if the force acting at joint D of the framework in problem 9.9 were directed along DB to the left?

[195 kN.]

9.11. A rigid horizontal beam 2 m long is supported by three vertical steel tie-rods each 500 mm long of cross-sectional area 600 mm². The bars are pinned to the beam at each end and at the mid-point. What vertically downward load could be applied to the beam (a) at one of the end pins, and (b) at the centre pin without causing the system to come within a factor of 2 of collapse? ($f_{yp} = 450$ MN/m².)

[(a) 202·5 kN. (b) 405 kN.]

9.12. Four steel bars are pinned together to form a square $ABCD$ (Fig. 9.9). In order to make the frame rigid a diagonal bar BD is then fitted. What is the maximum tensile load that can be applied to the frame along the line of diagonal AC without causing plastic failure? The cross-sectional area of bar BD is 10^{-3} m² and the areas of the

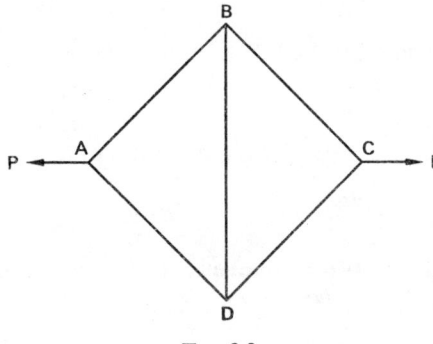

Fig. 9.9.

other members are such that all bars in the frame are equally strong. What is the area of the side bars? ($f_{yp} = 450$ MN/m².)

[450 kN. 0.7071×10^{-3} m².]

9.13. For the frame in problem 9.12 what would be the maximum load applied as before if a diagonal member were also fitted between A and C but not joined in any way to diagonal BD?

[900 kN.]

9.14. Two plates are butt-jointed and fastened by two cover plates with two lines each of four, 5 mm-diameter rivets on each side of the joint. Calculate the maximum safe load, using a safety factor of 3, which the joint can carry without failure by yielding of the rivets in shear. ($q_y = 140$ MN/m².)

[14.65 kN.]

9.15. A cylindrical steel shaft of 30 mm diameter has an axial torque applied until it yields to a depth of 5 mm below the surface. The yield stress of the material is 185 MN/m² and $G = 80$ GN/m². Find the total torque it carries and the angle through which it twists in a length of 1 m.

[1212·5 N m. 13·2°.]

9.16. A cylindrical steel shaft 25 mm diameter is subjected to an axial torque until the twist per metre is 15°. If the yield stress of the material in shear is 185 MN/m² and $G = 80$ GN/m², what is the torque and to what depth below the surface has yielding taken place?

[1284 N m. 3·6 mm.]

9.17. A solid shaft 40 mm diameter is made of steel, the shearing yield stress of which is 150 MN/m². After yielding, the stress remains constant for a very considerable increase of strain. Up to yield point $G = 80$ GN/m². If the shaft length is 600 mm calculate (a) the angle of twist and the torque when yielding begins, and (b) the torque needed to produce an angle of twist twice that in (a).

[London Univ., B.Sc. II]

[(a) 3·22°; 1882 N m. (b) 2436 N m.]

9.18. Find the first moment of area about the neutral axis for plastic bending S for a symmetrical I-section overall depth 100 mm, overall width 100 mm, with web and flanges 10 mm thick. What is the value of the shape factor $v \, (= S/Z)$?

[$S = 10\cdot 6 \times 10^{-5}$. $v = 1\cdot 18$.]

9.19. Find the plastic section modulus S, the elastic section modulus Z, and the shape factor $v = S/Z$ for a circular cross-section beam 20 mm diameter. (Note centroid of a semicircular lamina is at $4r/3\pi$ from the centre.)

$$\left[S = \frac{d^3}{6}. \quad Z = \frac{\pi d^3}{32}. \quad v = \frac{16}{3\pi} = 1\cdot 70. \right]$$

ELEMENTARY PLASTIC STRESS ANALYSIS

9.20. Show (a) that for a simply-supported beam plastic collapse will occur when the plastic moment of resistance $M_p = (nwL^2)/8$, where w is the maximum uniformly distributed load permitted over the span and n is the load factor, and (b) that for a built-in beam collapse will occur when $M_p = (nwL^2)/16$.

9.21. An I-section standard beam has a span of 6 m and a plastic modulus S of $1 \cdot 07 \times 10^{-3}$ m². If the yield stress of the material is 235 MN/m², find the uniformly distributed load which the beam can support if a load factor of 1·75 is applied and the beam is built-in at both ends.

[64·3 kN/m.]

9.22. What uniformly distributed load would the beam of problem 9.21 support with the same load factor if simply supported?

[32·1 kN/m.]

CHAPTER 10

Analysis of Stress in Engineering Components

Definitions and Theory

(a) The components chosen for study in this chapter have three things in common; they are fairly simple to analyse; some or all are included in most syllabuses; they illustrate the application of the basic theories developed in earlier chapters.

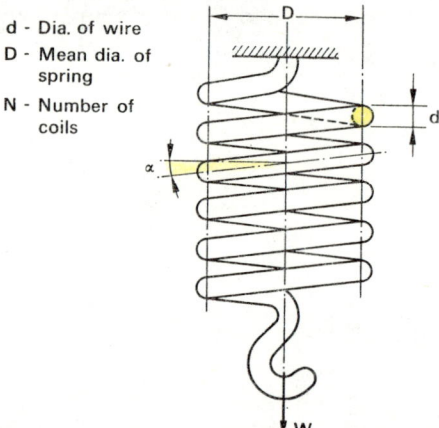

FIG. 10.1. Close-coiled spring.

(b) *Close-coiled helical springs of circular section wire.* A close-coiled spring is one whose helix angle α is small enough for the approximations $\sin \alpha \to 0$, $\cos \alpha \to 1$ to be acceptable. For such a spring (Fig. 10.1), an axial load W will cause a torque at every cross-section of the spring of

$$T = WD/2. \tag{1}$$

ANALYSIS OF STRESS

The work done by the load when deflecting the spring an axial distance δ will be stored as strain energy in the spring. The torque T acts about a horizontal axis and can be split into two components, $T\cos\alpha$ acting about the centre line of the spring wire and causing the cross-sections of the wire to twist through an angle θ relative to one another, and $T\sin\alpha$ acting perpendicular to the wire centre line and causing the wire to bend to an increased curvature and increase the axial twist, through an angle φ. Since α is small, $T\sin\alpha$ can be neglected and $T\cos\alpha \doteqdot T$. Hence the work done by the load is all stored as strain energy in twisting the wire,

i.e. $$\tfrac{1}{2}W\delta = \tfrac{1}{2}T\theta. \tag{2}$$

At any cross-section where the torque is T the maximum stress is given by

$$\frac{q}{d/2} = \frac{T}{J}. \tag{3}$$

Combining eqns. (1) to (3) it can be shown that under an axial load the spring stiffness W/δ is

$$\frac{W}{\delta} = \frac{Gd^4}{8D^3N},$$

maximum shear stress is $\quad q_M = \dfrac{8WD}{\pi d^3},$

and total strain energy is $\quad U = \dfrac{q_M^2}{4G} \times \text{volume of wire}.$

The other common form of loading is a torque M applied about the spring axis. The torque resolves into $M\cos\alpha$ about an axis perpendicular to the wire centre line and $M\sin\alpha$ about the centre line. The approximations $M\cos\alpha = M$; $M\sin\alpha = 0$ are made, and hence the applied torque effectively only causes the helix to tighten by twisting axially through an angle φ due to increased curvature of the wire. Hence work done = strain energy in bending.

$$\frac{1}{2}M\varphi = \frac{M^2 L}{2EI} \quad \text{and} \quad \varphi = \frac{ML}{EI}.$$

Assuming that the ratio D/d is large enough for the bending stress distribution to approximate to the simple bending relation,

$$f = \frac{My}{I} \quad \text{and} \quad f_M = \frac{32M}{\pi d^3}.$$

Strain energy $\quad U = \dfrac{f_M^2}{8E} \times \text{volume of spring}.$

(c) *Open-coiled helical springs of circular section wire.* Here α is large enough for the previous approximations to be invalid, and the bending and twisting effects of any load must be considered. Referring to Fig. 10.2, if the loading causes a torque T at a section such as A,

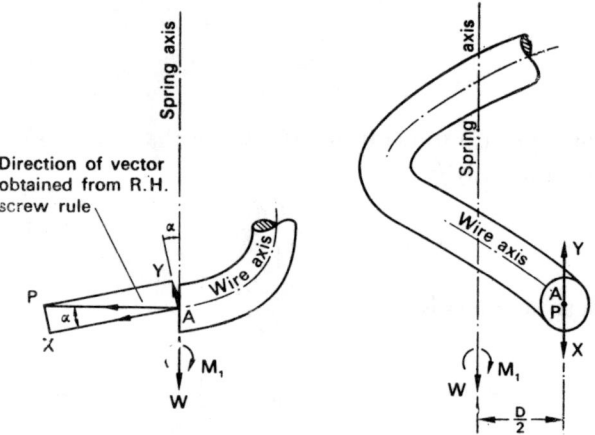

Fig. 10.2. Open-coiled spring.

which acts about an axis perpendicular to the section and can be represented by the vector **AP** then the vector can be resolved into perpendicular components. **AY** perpendicular to the centre line of the wire and in a plane parallel with the spring axis will be equal to $T \sin \alpha$ and will cause pure bending of the wire which has the effect of making the free end of the wire rotate round the spring axis through an angle φ. **AX** along the centre line of the wire and in a plane parallel with the

ANALYSIS OF STRESS

spring axis will be equal to $T \cos \alpha$ and will cause pure torsion of the wire causing a cross-section at the free end to rotate through an angle θ relative to a cross-section at the fixed end.

If an axial load W is applied, $T = WD/2$.

If a torque M_1 is applied about the axis of the spring it will cause a bending moment in the wire $M_1 \cos \alpha$ and a twisting effect about the wire centre line $M_1 \sin \alpha$. In the general case, where both axial load and axial torque are applied, as in Fig. 10.2;

Bending effect $= M_1 \cos \alpha - \dfrac{WD}{2} \sin \alpha$.

Twisting effect $= \dfrac{WD}{2} \cos \alpha + M_1 \sin \alpha$.

Strain energy $U = \dfrac{(M_1 \cos \alpha - WD/2 \sin \alpha)^2 L}{2EI}$

$+ \dfrac{(WD/2 \cos \alpha + M_1 \sin \alpha)^2 L}{2GJ}$.

In order to find the axial deflection δ, take $\partial U/\partial W$, and to find the axial twist φ, take $\partial U/\partial M_1$.

If only M_1 or W are acting it will be necessary to put M_1 or W equal to zero after partial differentiation. Usually M_1 and W will not be acting simultaneously.

Axial load only, $\delta = \dfrac{8WD^3N}{d^4 \cos \alpha} \cdot \left(\dfrac{\cos^2 \alpha}{G} + \dfrac{2 \sin^2 \alpha}{E} \right)$,

$\varphi = \dfrac{16WD^2N \sin \alpha}{d^4} \cdot \left(\dfrac{1}{G} - \dfrac{2}{E} \right)$.

Axial torque only, $\delta = \dfrac{16M_1D^2N \sin \alpha}{d^4} \cdot \left(\dfrac{1}{G} - \dfrac{2}{E} \right)$,

$\varphi = \dfrac{32M_1DN}{d^4 \cos \alpha} \cdot \left(\dfrac{\sin^2 \alpha}{G} + \dfrac{2 \cos^2 \alpha}{E} \right)$.

Bending stress can be found by assuming that the simple bending relation $f = My/I$ applies and shear stress by assuming that $q = (Td/2J)$

is valid (neither is in fact quite correct), where M and T are the total bending and twisting effects of whatever loads are acting.

To find the maximum direct and maximum shear stresses it must be remembered that both f and q are a maximum at the surface of the wire and must be added by complex stress system theory to find f_{max} and q_{max},

i.e. $\quad f_{max} = \dfrac{f}{2} + \sqrt{\left(\dfrac{f^2}{4} + q^2\right)} \quad$ and $\quad q_{max} = \dfrac{1}{2}\sqrt{(f^2 + 4q^2)}.$

(d) *Leaf springs* (Fig. 10.3). (i) Semi-elliptic type. The stack of leaves are assumed to be all of the same curvature and to move with no frictional effect between the leaves. With leaves of appropriate length

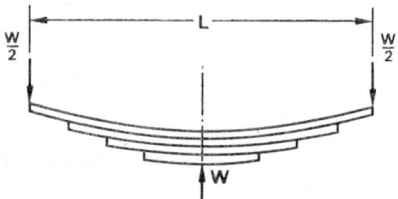

FIG. 10.3. Semi-elliptic leaf spring.

and tapered ends, the spring can be assumed to approximate to a rhomboidal simply-supported beam with a central point load. Making these assumptions the deflection δ, due to a central load W, will be

$$\delta = \frac{3WL^3}{8Nbt^3E},$$

where N is the number of leaves of breadth b, thickness t, and L is the distance between anchorages.

The bending stress is uniform throughout the length of each leaf and is the same for all leaves though it varies across the thickness.

$$f_{max} = \frac{3}{2}\frac{WL}{Nbt^2}.$$

(ii) Quarter-elliptic type,

$$\delta = \frac{6WL^3}{Nbt^3E},$$

$$f_{max} = \frac{6WL}{Nbt^2}.$$

In fact the friction between leaves and the impossibility of a real spring approximating closely to the theoretical shape means that a real spring may be 10–25 per cent stiffer than the above calculations would lead one to expect.

(e) *Thick cylinders.* In a previous chapter the case of pressurized thin cylinders was dealt with by assuming that stress in the radial direction could be neglected. It was then shown that the circumferential stress f_θ due to an internal pressure p was $f_\theta = (pr)/t$. For the thick cylinder shown in Fig. 10.4, not only is the radial stress f_r not negli-

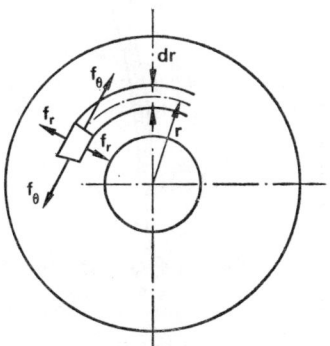

FIG. 10.4. Thick cylinders.

gible, but both it and f_θ may vary with the radial distance at which they act. A complete analysis involves the general equations of analysis of variation of stress and strain from which Lamé's equations can be derived, i.e. in a thick cylinder,

$$f_\theta = A + \frac{B}{r^2} \quad \text{and} \quad f_r = A - \frac{B}{r^2},$$

where f_θ and f_r are the circumferential and radial stresses at radius r, and A and B are constants to be determined from the boundary conditions of the problem.

(f) *Struts.* A strut is a member which carries compressive loading along, or parallel with its axis. If the length is much greater than the least transverse dimension ($L > 30D$) the strut will fail by buckling when the load exceeds a critical value which is determined by the properties of the strut material, the least second moment of area of the cross-section, the length and the type of end fixing.

For the long pin-ended strut (Fig. 10.5a),

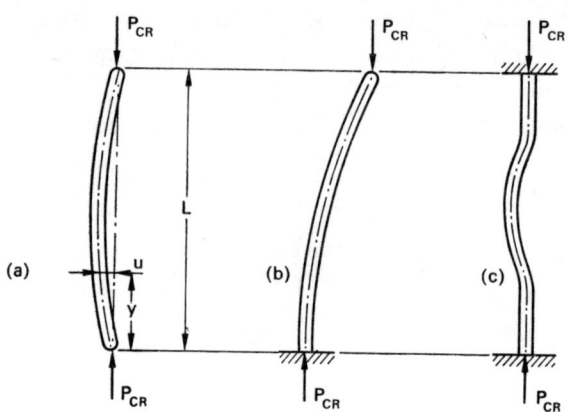

Fig. 10.5. Struts.

Euler's equation,

$$P_{cr} = \frac{\pi^2 EI}{L^2}$$

has been found to give a good estimate of the critical load P_{cr} at which the strut will buckle.

For a long strut with one fixed and one free end (Fig. 10.5b),

$$P_{cr} = \frac{\pi^2 EI}{4L^2}.$$

For a long strut with both ends fixed (Fig. 10.5c),

$$P_{cr} = \frac{4\pi^2 EI}{L^2}.$$

Euler's equation can be rewritten

$$P_{cr} = \frac{\pi^2 EAk^2}{L^2},$$

where A is the cross-sectional area and k the least radius of gyration, and dividing by A gives

$$\frac{P_{cr}}{A} = \frac{\pi^2 E}{(L/k)^2}.$$

P_{cr}/A represents compressive stress in the member due to direct loading and L/k is called the slenderness ratio. For a very short strut L/k is small and therefore P_{cr}/A would be very large, i.e. failure by buckling would not take place until after P_{cr}/A had exceeded the yield stress in compression and permanent deformation had taken place. In order to allow for this a number of empirical formulae have been derived which give results approximately the same as the Euler equation for long struts but which can be used for medium length and short struts as well. A typical formula is the Rankine–Gordon relation

$$P_{cr} = \frac{fA}{1+\alpha(L/k)^2},$$

where f is the stress at yield point and α is a constant that depends on the material of the strut and the end fixing conditions, in fact, the constant

$$\alpha = \frac{f}{\beta\pi^2 E},$$

where β is the numerical coefficient in the Euler formula, i.e. 1 for a pin-ended, 4 for a fixed-ended, and $\frac{1}{4}$ for one free- and one fixed-end, strut.

It should be noted that α is a dimensionless quantity if f and E are in the same units and that values quoted in existing sources can be used unchanged in SI.

(g) *Theories of failure.* The term "failure" in this context means failure by yielding and plastic deformation; the theories are therefore concerned with ductile materials and seek to establish criteria which will allow yield stress data from a uniaxial tensile test to be used in deciding whether a member will fail by yielding under any given condition of complex stressing.

The theories usually studied are listed in Table 10.1.

TABLE 10.1

Name	Yielding occurs when	Comment
Maximum principal stress theory	$f_1 = f_{yt}$ or $f_3 = f_{yc}$	Simple but only gives good agreement with experiment for uniaxial loading
Maximum principal strain theory	$f_1 - \nu(f_2+f_3) = f_{yt}$ or $f_3 - \nu(f_1+f_2) = f_{yc}$	Another early attempt at a theory, now seldom used
Maximum shear stress theory (Tresca criterion)	$f_1 - f_3 = f_{yt}$	Fair agreement with experiment and mathematically simple
Shear strain energy theory (von Mises criterion)	$(f_1-f_2)^2 + (f_2-f_3)^2 + (f_3-f_1)^2 = 2f_{yt}^2$	Best of the available theories for ductile materials

In Table 10.1, f_{yt} is the yield stress in uniaxial tension and f_{yc} is the yield stress in uniaxial compression.

(h) *Composite beams.* The general method of analysing beams of two materials was developed for wooden beams reinforced with a metal strip or strips. The method, with modifications, can be used for reinforced concrete with tension reinforcement, but beyond the simplest cases it is easier to treat the reinforced concrete beam as a separate problem.

General method: making the usual assumption that plane cross-sections before bending remain planes after bending, then the strain e_y at y from the neutral axis will be

$$e_y = \frac{y}{R},$$

where R is the radius of curvature of the neutral axis.

Since stress and strain are related by $f/e = E$ for the uniaxial case then for a beam where there are two materials A and B both arranged to be at y from the neutral axis, then

$$\frac{f_A}{E_A} = \frac{f_B}{E_B} = \frac{y}{R}.$$

Hence it can be shown that one of the materials in the beam, say A, can be removed and replaced by an additional piece of material B of the same depth but of breadth E_A/E_B times its actual breadth.

The stresses in the beam can then be found as if it were a homogeneous beam of material B. If the stress so found is carried in the real beam by material A, then the true stress will be E_A/E_B times the stress calculated for the imagined homogeneous beam.

Reinforced concrete: the assumptions made are:
(1) plane cross-sections remain plane;
(2) there is perfect adhesion between steel and concrete;
(3) concrete does not carry any tensile stress; all tension is carried by the reinforcement;
(4) the cross-sectional area of reinforcement is small enough for the stress in it to be constant.

Referring to Fig. 10.6, one can proceed as follows:
By similar triangles,

$$\frac{\text{Maximum strain in compression concrete}}{\text{Strain in reinforcement}} = \frac{h}{d-h}$$

$$\therefore \frac{f_c/E_c}{f_s/E_s} = \frac{h}{d-h}. \quad (1)$$

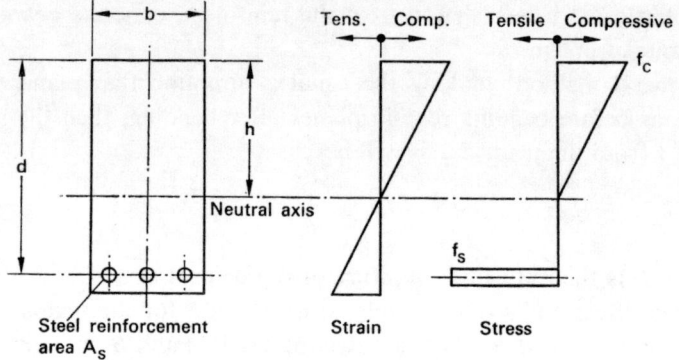

Fig. 10.6. Tension-reinforced concrete beam.

The total force across a cross-section is zero.
Therefore compressive force in concrete = tensile force in steel,

$$\tfrac{1}{2} f_c bh = f_s A_s. \tag{2}$$

Eliminating f_c/f_s between (1) and (2) gives

$$0 = bh^2 + \frac{2E_s}{E_c} A_s h - \frac{2E_s}{E_c} dA_s. \tag{3}$$

This equation depends only on the properties of the beam geometry and materials and locates the position of the neutral axis.

Since force in steel = force in concrete, the two form a couple which is the resisting moment of the beam,

i.e. $\quad M = \dfrac{f_c}{2} bh \left(d - \dfrac{h}{3}\right) \quad$ or $\quad M = f_s A_s \left(d - \dfrac{h}{3}\right). \tag{4}$

Notes

(1) If the beam has steel compression reinforcement, eqn. (1) is unaltered but eqn. (2) becomes

> Compressive force in concrete + compressive force in compression-reinforcement = tensile force in tension-reinforcement.

ANALYSIS OF STRESS

Equations (3) and (4) must be altered to conform.

(2) In order to use material most economically it is usual to design the beam so that f_s and f_c both reach predetermined values simultaneously, and if the depth of the beam d is known, h can be found from eqn. (1), A_s from (2), and M from (4). Equation (3) is redundant.

(3) It would be possible to solve problems by the general method; this would involve transforming the area of steel A_s into an imaginary area of concrete

$$\frac{E_s}{E_c} A_s.$$

The odd situation then arises that real concrete is imagined not to carry tension but that imaginary concrete does.

Worked Examples

10.1. A close-coiled helical spring made of steel rod 10 mm diameter has 15 turns of 70 mm mean diameter. The spring is loaded axially, the maximum load being 1000 N, and this load produces a deflection of 50 mm.

A second spring is to be made to deflect 50 mm under a load of 1500 N; the maximum shear stress, material, and ratio of wire diameter to coil diameter are to be the same for both springs. Find wire diameter, mean coil diameter, and the number of turns in the second spring. [London Univ., 1964]

Solution

From the three equations, $T = WD/2$, (1)

$$\tfrac{1}{2}W\delta = \tfrac{1}{2}T\theta, \tag{2}$$

$$\frac{q}{d/2} = \frac{T}{J} = \frac{G\theta}{L}, \tag{3}$$

combining eqn. (1) with the first part of eqn. (3), gives

$$\frac{q}{d/2} = \frac{WD/2}{J},$$

and substituting for J, $\quad W = \dfrac{\pi d^3 q}{8D}.$

Combining all three gives

$$\delta = \frac{\pi D^2 N q}{G d}.$$

Using suffix 1 for quantities referring to spring 1 and suffix 2 for quantities referring to spring 2,

$$\delta_1 = \delta_2.$$

$$\therefore \frac{\pi D_1^2 N_1 q_1}{G_1 d_1} = \frac{\pi D_2^2 N_2 q_2}{G_2 d_2}.$$

Now $\quad \dfrac{D_1}{d_1} = \dfrac{D_2}{d_2} \quad$ and $\quad q_1 = q_2.$

Also, $\quad G_1 = G_2.$

$$\therefore D_1 N_1 = D_2 N_2.$$

The maximum load for the two springs produces the same maximum shearing stress which can be called q.

$$\therefore 1000 = \frac{\pi d_1^3 q}{8 D_1} \quad \text{for spring 1,}$$

and $\quad 1500 = \dfrac{\pi d_2^3 q}{8 D_2} \quad$ for spring 2.

Dividing these equations gives

$$\frac{2}{3} = \frac{d_1^3}{D_1} \times \frac{D_2}{d_2^3}.$$

ANALYSIS OF STRESS

But
$$\frac{d_1}{D_1} = \frac{d_2}{D_2}.$$

$$\therefore \frac{2}{3} = \frac{d_1^2}{d_2^2},$$

$$d_2 = \sqrt{(\tfrac{3}{2})}\, d_1$$
$$= 10\sqrt{(\tfrac{3}{2})} \text{ mm}$$
$$= 12\cdot 25 \text{ mm},$$

and
$$D_2 = \frac{d_2}{d_1} D_1$$
$$= 85\cdot 75 \text{ mm}.$$

Also,
$$N_2 = \frac{D_1}{D_2} N_1 = \sqrt{\left(\frac{2}{3}\right)} \times 15 = 12\cdot 25 \text{ turns.}$$

10.2. Two close-coiled helical springs stand on the same horizontal surface one inside the other. The particulars of the two springs are: outer spring—mean diameter 44 mm, wire diameter 4·85 mm, number of turns 10, free length 100 mm: inner spring—mean diameter 31·5 mm, wire diameter 4·06 mm, number of turns 8, free length 82 mm.

A load W in the form of a flat plate weighing 500 N is placed so that it rests horizontally on the outer spring and compresses it until some load is also taken by the inner spring. For each spring, calculate (a) change in length, (b) load carried, (c) maximum shearing stress. ($G = 80$ GN/m².) [London Univ.]

Solution

Substituting given values in

$$\frac{W}{\delta} = \frac{Gd^4}{8D^3N},$$

the stiffness of the outer spring is found to be 6·5 kN/m and of the inner spring 10·85 kN/m.

Load required to compress outer spring to same length as the inner $= 6.5 \times 10^3 \times (100-82) \times 10^{-3}$

$= 117$ N.

The remainder of the load $= 500 - 117$
$= 383$ N,

will compress both springs that have a combined stiffness of 17·35 kN/m.

Both springs together compress $383/17350$ m $= 22.1$ mm.

Therefore outer spring compresses 40·1 mm and inner spring compresses 22·1 mm.

Outer spring carries $6.5 \times 10^3 \times 40.1 \times 10^{-3} = 260$ N, and inner spring carries $10.85 \times 10^3 \times 22.1 \times 10^{-3} = 240$ N.

Maximum shearing stress $= \dfrac{8WD}{\pi d^3}$.

Therefore for outer spring,

$$q_M = \frac{8 \times 260 \times 44 \times 10^{-3}}{\pi \times 4.85^3 \times 10^{-9}}$$

$= 255$ MN/m²

and for inner spring,

$$q_M = \frac{8 \times 240 \times 31.5 \times 10^{-3}}{\pi \times 4.06^3 \times 10^{-9}}$$

$= 288$ MN/m².

10.3. An open-coiled helical spring is made of steel wire 7·6 mm diameter; the coils have a mean diameter of 64 mm and a pitch of 50 mm. If the spring carries an axial load of 450 N, find the maximum shearing stress produced and the number of turns required so that the extension caused by the load is 26 mm. ($E = 200$ GN/m² and $G = 78.2$ GN/m².) [London Univ.]

Solution

The maximum shear stress will be that which occurs on a certain plane at the surface of the wire where a stress f due to the bending effect combines with the stress q due to the twisting effect.

Hence,
$$f = \frac{WD/2 \; d/2 \sin \alpha}{\pi d^4/64}$$

$$= \frac{16WD \sin \alpha}{\pi d^3}$$

and
$$q = \frac{WD/2 \cos \alpha \; d/2}{\pi d^4/32}$$

$$= \frac{8WD \cos \alpha}{\pi d^3}.$$

In a complex stress system,

$$q_{max} = \frac{1}{2} \sqrt{(f^2 + 4q^2)}$$

$$= \frac{1}{2} \sqrt{\left[\left(\frac{16WD \sin \alpha}{\pi d^3}\right)^2 + 4\left(\frac{8WD \cos \alpha}{\pi d^3}\right)^2\right]}$$

$$= \frac{8WD}{\pi d^3} \sqrt{(\sin^2 \alpha + \cos^2 \alpha)}$$

$$= \frac{8WD}{\pi d^3}$$

$$= \frac{8 \times 450 \times 64 \times 10^{-3}}{\pi \times 7 \cdot 6^3 \times 10^{-9}} \; \text{N/m}^2$$

$$= 167 \; \text{MN/m}^2.$$

The number of coils N can be found by substituting in the expression

$$\delta = \frac{8WD^3N}{d^4 \cos \alpha}\left(\frac{\cos^2 \alpha}{G} + \frac{2 \sin^2 \alpha}{E}\right)$$

developed in the initial section of the chapter where

$$\alpha = \tan^{-1}\left(\frac{\text{half pitch}}{\text{coil diameter}}\right)$$

$$= \tan^{-1}\frac{25}{64}$$

$$= 21\cdot 35°.$$

Hence $N = 7$ turns.

10.4. Derive an expression for the axial stiffness of an open-coiled helical spring made of wound wire of diameter d, mean coil diameter D, helix angle α. Calculate the percentage error in the value obtained for the stiffness if the inclination of the coils is neglected for a spring in which $\alpha = 30°$. ($E = 2\cdot 5\ G$.) [London Univ., B.Sc. II]

Solution

Axial load W will cause a twisting torque $(WD/2)\cos\alpha$ and a bending effect $(WD/2)\sin\alpha$.

$$\text{Strain energy} = \left(\frac{WD}{2}\sin\alpha\right)^2\frac{\pi DN}{2EI\cos\alpha} + \left(\frac{WD}{2}\cos\alpha\right)^2\frac{\pi DN}{2GJ\cos\alpha},$$

$$U = \frac{W^2\pi D^3 N}{8\cos\alpha}\left[\frac{\sin^2\alpha}{E\pi d^4/64} + \frac{\cos^2\alpha}{G\pi d^4/32}\right],$$

$$U = \frac{4W^2 D^3 N}{d^4 \cos\alpha}\left[\frac{2\sin^2\alpha}{E} + \frac{\cos^2\alpha}{G}\right].$$

Axial deflection

$$= \frac{\partial U}{\partial W}$$

$$= \delta.$$

$$\therefore\ \delta = \frac{8WD^3 N}{d^4 \cos\alpha}\left[\frac{\cos^2\alpha}{G} + \frac{2\sin^2\alpha}{E}\right].$$

$$\text{Stiffness} = \frac{W}{\delta} = \frac{d^4 GE\cos\alpha}{8D^3 N(E\cos^2\alpha + 2G\sin^2\alpha)}.$$

ANALYSIS OF STRESS

If the inclination of the coils is neglected

i.e. $\alpha \doteq 0$, then $\cos \alpha = 1$,

$$\sin \alpha = 0,$$

$$\frac{W}{\delta} = \frac{d^4 GE}{8D^3 NE}$$

$$= \frac{Gd^4}{8D^3 N},$$

which is the expression for a close-coiled spring.

If the calculated stiffness using the close-coiled expression is S_c and the true stiffness using the open-coiled expression is S_o,

$$\text{error} = \frac{S_c - S_o}{S_o} \times 100 \text{ per cent.}$$

Substituting the value $\alpha = 30°$ and putting $E = 2.5\ G$ in the expression for S_o gives, after cancellation,

$$\text{error} = \frac{1 - \dfrac{2.5 \times 0.8660}{2.5 \times 0.75 + 0.5}}{\dfrac{2.5 \times 0.8660}{2.5 \times 0.75 + 0.5}} \times 100 \text{ per cent}$$

$$= (1.097 - 1) \times 100 \text{ per cent}$$

$$= 9.7 \text{ per cent.}$$

10.5. A semi-elliptic steel spring has a span of 760 mm and is required to carry a proof load of 7·5 kN and the central deflection is not to exceed 50 mm. The stress due to bending must not exceed 390 MN/m². If the ratio of breadth to thickness $= 12$ and $E = 207$ GN/m², find the number, breadth, and thickness of plates, the actual deflection under the proof load, and the initial radius of the spring. Assume plates are only available in "whole millimetre" thicknesses.

Solution

N.B.—Proof load is load required to straighten spring.

$$\delta = \frac{3WL^3}{8Nbt^3E} \quad (1)$$

and

$$f_{max} = \frac{3WL}{2Nbt^2}. \quad (2)$$

Dividing eqn. (1) by eqn. (2) gives

$$\frac{\delta}{f_{max}} = \frac{L^2}{4tE},$$

$$t = \frac{L^2 f_{max}}{4\delta E},$$

$$t = \frac{0\cdot76^2 \times 390 \times 10^6}{4 \times 50 \times 10^{-3} \times 207 \times 10^9}$$

$$= 5\cdot45 \text{ mm},$$

$$= 6 \text{ mm, to nearest millimetre.}$$

Transposing eqn. (2),

$$N = \frac{3}{2} \frac{WL}{f_{max}dt}.$$

Substituting, using $t = 6$ mm,

$$N = 8\cdot5$$

$$= 9 \text{ to next highest whole number.}$$

Actual deflection from eqn. (1) using $N = 9$, $t = 6$ mm, is

$$\delta = 42\cdot8 \text{ mm}.$$

The spring is initially curved to an arc of radius R_0 and straightens to become a chord of length 760 mm by deflecting 42·8 mm. From geometry of chord and segment,

$$R_0 = \frac{L^2}{8\delta}$$

$$= 1695 \text{ mm}.$$

10.6. A thick cylinder with closed ends has an inside diameter of 200 mm and an outside diameter of 400 mm. If the internal pressure is 10 MN/m² and the external pressure is zero, find the values of radial, hoop and axial stress at the inside of the curved walls.

Solution

From Lamé's equations,

$$f_r = A - \frac{B}{r^2} \quad \text{and} \quad f_\theta = A + \frac{B}{r^2}.$$

The first step is to find constants A and B by putting in boundary conditions, i.e. values of r for which f_r or f_θ are known.

In this problem, at $r = 0.1$ m, f_r = internal pressure = -10^7 N/m².

N.B.—The negative sign appears since the pressure will give rise to a compressive stress.

At $r = 0.2$ m, f_r = external pressure = 0.

$$\therefore \quad 0 = A - \frac{B}{0.04}$$

$$-10^7 = A - \frac{B}{0.01}.$$

$$\therefore \quad A = \frac{100B}{4},$$

$$B = \frac{4}{3} \times 10^5, \quad A = \frac{10^7}{3}.$$

At $r = 0.1$ m, i.e. at the inside of the walls,

$$f_r = \frac{10^7}{3} - \frac{4}{3} \times \frac{10^5}{0.01}$$

$$= -10^7 \text{ N/m}^2$$

(which was already known),

$$f_\theta = A + \frac{B}{r^2}$$

$$= \frac{10^7}{3} + \frac{4}{3} \times \frac{10^5}{0.01} = \frac{5}{3} \times 10^7 \text{ N/m}^2.$$

Since the end of the cylinder is closed there will be a longitudinal stress f_L which is assumed evenly distributed over the cross-section. Therefore for all values of r,

$$f_L = \frac{p\pi r_i^2}{(r_0^2 - r_i^2)},$$

where r_i is the inner radius and r_0 is the outer radius,

$$= \frac{1}{3} \times 10^7 \text{ N/m}^2.$$

10.7. A straight bar of steel 1150 mm long and 21 mm by 7 mm rectangular cross-section is loaded axially in compression until it buckles elastically.

Assuming Euler's formula for a pin-ended strut applies, find the maximum central deflection before the material reaches its yield stress of 360 MN/m². ($E = 200$ GN/m².) [London Univ.]

Solution

Euler's formula for pin-ended strut is

$$P_{cr} = \frac{\pi^2 EI}{L^2}. \tag{1}$$

Strut will buckle in plane of least cross-sectional dimension, i.e.

$$I = \frac{21 \times 7^3}{12} \times 10^{-12} \text{ m}^4.$$

$$= 6 \times 10^{-10} \text{ m}^4.$$

Substituting in eqn. (1),

$$P_{cr} = 894 \text{ N}.$$

In Fig. 10.7 it can be seen that the middle of the strut, which has moved laterally, relative to the line of the load, through a distance δ will be carrying a direct stress P_{cr}/A together with a bending moment $P_{cr}\,\delta$.

ANALYSIS OF STRESS

Therefore total compressive stress (on inside of bend)

$$= \frac{P_{cr}}{A} + \frac{P_{cr}\,\delta y}{I},$$

where

$$y = \tfrac{7}{2} \times 10^{-3} \text{ m},$$
$$I = 6 \times 10^{-10} \text{ m}^4.$$

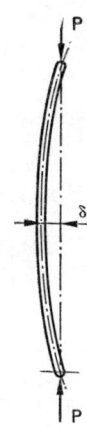

Fig. 10.7.

When compressive stress = yield stress,

$$360 \times 10^6 = 894(147 \times 10^{-6} \times \delta \times \tfrac{7}{12} \times 10^7)$$
$$\therefore \delta = 67\cdot 9 \text{ mm}.$$

10.8. Starting from the Euler formula for the buckling load of a pin-ended slender strut,

$$P_{cr} = \frac{\pi^2 EI}{L^2},$$

(a) deduce a similar relation for a strut with two fixed ends.

(b) For the strut in part (a) find a relation between the compressive stress due to direct loading under critical conditions P_{cr}/A and the slenderness ratio L/k.

(c) Sketch a graph of P_{cr}/A against $(L/k)^2$ and explain the limitations on the validity of the Euler relation for a short strut.

Solution

(a) Comparing Fig. 10.5a and c it can be seen that the shape of the fixed-ended strut is such that there are two points of contraflexure at the quarter points. Thus the middle half will be exactly similar in deformed shape to the whole length of the pin-ended strut. The critical load for the fixed strut will be given by the same relation if $L/2$ is substituted for L,

i.e.
$$P_{cr} = \frac{\pi^2 EI}{(L/2)^2}$$
$$= \frac{4\pi^2 EI}{L^2}.$$

Note that the same relation is obtained from first principles by setting up the differential equation for deflection,

$$\frac{d^2 u}{dy^2} + k^2 u = 0, \quad \text{where} \quad k^2 = \frac{P}{EI},$$

which has the general solution

$$u = A \cos ky + B \sin ky.$$

For the fixed-ended strut the slope $du/dy = 0$ at $y = 0$ and at $y = L$,

i.e. $\quad \dfrac{du}{dy} = -Ak \sin ky + Bk \cos ky;$

at $\quad y = 0, \quad \dfrac{du}{dy} = 0 \quad$ only if $\quad B = 0;$

at $\quad y = L, \quad \dfrac{du}{dy} = 0 \quad$ only if $\quad \sin kL = 0;$

i.e. $\quad k = \dfrac{2\pi}{L}, \dfrac{4\pi}{L},$ etc.

ANALYSIS OF STRESS

$$\therefore \quad k^2 = \frac{4\pi^2}{L^2}.$$

$$\therefore \quad \frac{P_{cr}}{EI} = \frac{4\pi^2}{L^2},$$

$$P_{cr} = \frac{4\pi^2 EI}{L^2}.$$

(b) Direct compressive stress $= P_{cr}/A = f_{cr}$.
Therefore dividing by A and substituting Ak^2 for I,

$$P_{cr}/A = \frac{4\pi^2 EAk^2}{AL^2},$$

$$f_{cr} = \frac{4\pi^2 E}{(L/k)^2}.$$

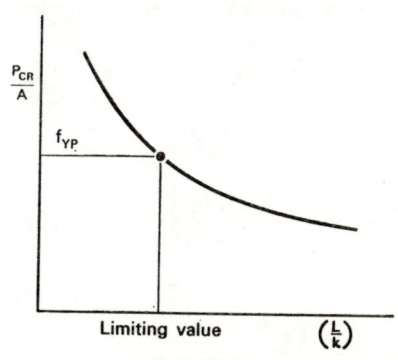

Fig. 10.8.

(c) Figure 10.8 shows how P_{cr}/A will become very large as $(L/k)^2$ becomes small. If P_{cr}/A becomes large enough, the direct stress will exceed the compressive yield stress and yielding will take place. The Euler relation was derived for elastic conditions and therefore cannot be used below:

$$f_{yp} = \frac{4\pi^2 E}{(L/k)^2}$$

i.e.
$$(L/k)^2 = \frac{4\pi^2 E}{f_{yp}}.$$

248 STRESS ANALYSIS PROBLEMS IN S.I. UNITS

Taking $E = 200$ GN/m², $f_{yp} = 200$ MN/m², $\pi^2 \doteqdot 10$.
The limiting value of L/k will be $\sqrt{(4 \times 10^4)} = 200$,
below which the Euler result will be completely invalid.

(N.B.—There will be a region of values of L/k just above the limit where the Euler relation will begin to show departures from experimental results.)

10.9. A mild steel shaft of 50 mm diameter carries a bending moment of 2000 N m. If the yield point stress in simple tension is 200 MN/m², find the maximum torque that can be applied in addition to the bending moment without causing yielding, using each of the following three theories of failure: (a) maximum principal stress, (b) maximum shear stress, and (c) maximum shear strain energy (von Mises).

Solution

The principal stresses $f_{1,2}$ are given by

$$f_{1,2} = \frac{f_x + f_y}{2} \pm \sqrt{\left[\frac{1}{4}(f_x - f_y)^2 + q_{xy}^2\right]}.$$

Here $\qquad f_y = 0$,

and bending stress, $f_x = \dfrac{My}{I}$

$= \dfrac{32M}{\pi d^3}$

$= 163 \cdot 5$ MN/m²,

$q_{xy} =$ torsion shear stress

$= \dfrac{16T}{\pi d^3}$

$= 40 \cdot 8T$ kN/m²,

where T is the torque.

ANALYSIS OF STRESS

(a) Maximum principal stress theory is that yielding takes place if $f_1 = f_{yt}$,

i.e., $200 \times 10^6 = \dfrac{163.5 \times 10^6 - 0}{2}$

$$+ \sqrt{\left[\dfrac{1}{4}(163.5 \times 10^6)^2 + (40.8 T \times 10^3)^2\right]}.$$

$\therefore T = 2092$ Nm.

(b) Maximum shear stress theory is that yielding takes place if $f_1 - f_3 = f_{yt}$.

Hence, f_3 = maximum compressive stress

$$= \dfrac{f_x}{2} - \sqrt{\left[\left(\dfrac{f_x}{2}\right)^2 + q_{xy}^2\right]},$$

$$f_1 = \dfrac{f_x}{2} + \sqrt{\left[\left(\dfrac{f_x}{2}\right)^2 + q_{xy}^2\right]}$$

$$f_1 - f_3 = 2\sqrt{\left[\left(\dfrac{f_x}{2}\right)^2 + q_{xy}^2\right]},$$

$200 \times 10^6 = 2\sqrt{[(81.75 \times 10^6)^2 + (40.8 \times 10^3 T)^2]}$,

$T = 1410$ N m.

(c) von Mises criterion gives,

$$(f_1 - f_2)^2 + (f_2 - f_3)^2 + (f_3 - f_1)^2 = 2f_{yt}^2.$$

Here $f_2 = 0.$

$\therefore 2f_1^2 - 2f_1 f_3 + 2f_3^2 = 2f_{yt}^2,$

$f_1^2 - f_1 f_3 + f_3^2 = f_{yt}^2.$

Substituting the values for f_1 and f_3 and simplifying gives

$$f_x^2 + 3q_{xy}^2 = f_{yt}^2$$

$(163.5 \times 10^6)^2 + 3(40.8 \times 10^3 T)^2 = (200 \times 10^6)^2.$

$\therefore T = 1623$ N m.

10.10. Two rectangular bars—one brass and the other steel—each 38 mm wide and 9·5 mm thick, are placed together and firmly fastened along their length to form a compound bar 38 mm wide and 19 mm deep. The compound bar is simply supported over a span of 750 mm with the brass on top. What is the maximum load which can be applied in the centre without the stress in the steel exceeding 100 MN/m² or the stress in the brass exceeding 70 MN/m²? ($E_{brass} = 86$ GN/m². $E_{steel} = 200$ GN/m².) [London Univ.]

Solution

A homogeneous beam could be substituted for the compound beam if it were all steel provided that the brass were replaced by a portion of steel of the same depth but of breadth

$$\frac{E_B}{E_s} \times 38 \text{ mm} = \frac{86}{200} \times 38 = 16\cdot34 \text{ mm}.$$

Referring to Fig. 10.9, where the resulting steel section is illustrated, the distance $\bar{y}$ of the neutral axis from the base is found to be $\bar{y} = 7\cdot6$ mm,

while $I_{NA} = 1\cdot366 \times 10^{-8}$ m⁴.

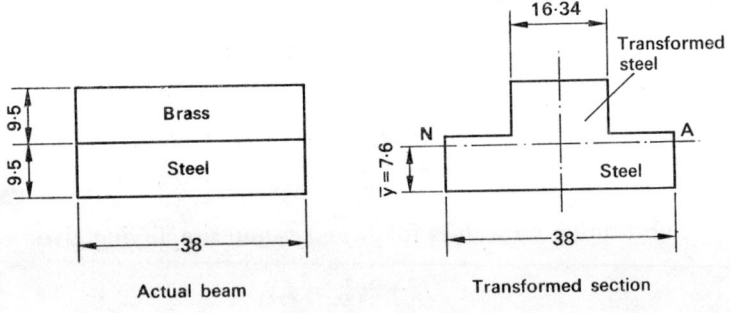

Actual beam   Transformed section

Dimensions in mm

Fig. 10.9.

ANALYSIS OF STRESS

Maximum stress in steel will occur at bottom of section where

$$f = \frac{M \times 7 \cdot 6 \times 10^{-3}}{1 \cdot 366 \times 10^{-8}},$$

but
$$M = \frac{WL}{4}$$
$$= \frac{0 \cdot 75W}{4},$$

and f is not to exceed 100 MN/m²,

$$100 \times 10^6 = \frac{0 \cdot 75W \times 7 \cdot 6 \times 10^{-3}}{4 \times 1 \cdot 366 \times 10^{-8}},$$

i.e. $\quad W = 960$ N.

Maximum stress in brass will occur at top of section. If section were all steel, stress would be

$$f = \frac{M \times 11 \cdot 4 \times 10^{-3}}{1 \cdot 366 \times 10^{-8}}.$$

Since the material is in fact brass,

$$f = \frac{E_B}{E_s} \frac{M \times 11 \cdot 4 \times 10^{-3}}{1 \cdot 366 \times 10^{-8}},$$

and f is not to exceed 70 MN/m²,

$$70 \times 10^6 = \frac{86}{200} \times \frac{0 \cdot 75W \times 11 \cdot 4 \times 10^{-3}}{4 \times 1 \cdot 366 \times 10^{-8}},$$

i.e. $\quad W = 1085$ N,

i.e. stress in steel is limiting factor and maximum load = 960 N.

10.11. A concrete beam 300 mm wide and 460 mm deep has four reinforcing rods each of 3 cm² area centred 60 mm above the tension face. If the ratio of the moduli of steel to concrete is 16 : 1 find the position of the neutral axis. What will be the maximum stresses in steel and concrete when the beam is used simply supported over a span of 3·6 m to carry a central point load of 55 kN?

Solution

The neutral axis can be located by solving the equation

$$bh^2 + 2\frac{E_s}{E_c} A_s h - 2\frac{E_s}{E_c} dA_s = 0,$$

where $b = 0·3$, $\dfrac{E_s}{E_c} = 16$, $A_s = 4\times 3\times 10^{-4}$, $d = 0·4$,

giving
$$h = 0·171 \text{ m}.$$

Stress in concrete f_c is obtained from

$$M = \frac{f_c}{2} bh \left(d - \frac{h}{3}\right),$$

where
$$M = \frac{WL}{4} = \frac{55\times 10^3 \times 3·6}{4}.$$

$$\therefore f_c = 5·62 \text{ MN/m}^2.$$

Stress in steel f_s is obtained from

force in steel = force in concrete,

$$f_s A_s = \frac{f_c}{2} bh,$$

i.e.
$$f_s = \frac{5·62\times 10^6 \times 0·3\times 0·171}{2\times 12\times 10^{-4}} \text{ N/m}^2,$$

$$f_s = 120 \text{ MN/m}^2.$$

Problems

10.12. A close-coiled helical spring is required to carry an axial load of 2·7 kN at a maximum shearing stress of 345 MN/m². The mean coil diameter is to be 8 times the diameter of the steel wire. Find the

ANALYSIS OF STRESS 253

wire and coil diameters. If $G = 80 \times 10^9$ N/m² and there are ten coils, find the extension of the spring when the load of 2·7 kN is applied.

[London Univ., 1951]

[$d = 12·6$ mm. $D = 100·8$ mm. $\delta = 110$ mm.]

10.13. A close-coiled spring has fourteen free coils of 16·5 mm diameter and is made of 2 mm diameter wire. Its original length is 60 mm. What compressive load will reduce the length to 44 mm? What will be the maximum stress in the wire at this load? ($G = 76$ GN/m².)

[London Univ., 1945]

[38·7 N. 203·5 MN/m².]

10.14. Wire of circular section 7·5 mm diameter is used to make a close-coiled helical spring whose stiffness is 21 kN/m. The maximum static load the spring is to carry is 250 N and the maximum allowable stress in the wire is 85 MN/m². If $G = 83$ GN/m², find the mean diameter of the coil and the number of turns required.

[London Univ., B.Sc. I]

[56·25 mm. 8·8 turns.]

10.15. A spring is made with 100 turns of 5 mm diameter steel wire close-coiled to a mean diameter of 50 mm. If $G = 80$ GN/m² find the spring stiffness.

[500 N/m.]

10.16. An open-coiled spring is made with ten turns of 10 mm diameter wire wound at an angle $\alpha = 30°$ to a mean coil diameter of 114 mm. Find (a) the axial extension δ, and (b) the angle of twist φ of the free end about the coil axis when an axial load of 90 N is applied. ($E = 207$ GN/m² and $G = 83$ GN/m².)

[London Univ.]

[$\delta = 14·1$ mm. $\varphi = 1·28°$.]

10.17. In an open-coiled helical spring of 10 coils, the stresses due to bending and to twisting are respectively 96 MN/m² and 103 MN/m² when the spring carries a certain axial load. If the coil diameter is eight times the wire diameter, find the value of the axial load and the diameter of the wire for a maximum extension of 18 mm. ($E = 207$ GN/m² and $G = 76$ GN/m².) [London Univ.]

[181 N. 5·7 mm.]

(N.B.—α can be obtained by writing the equations for f and q, eliminating W and substituting the given values for f and q; $\alpha = 25°$.)

10.18. An open-coiled helical spring made of steel rod 13 mm diameter has ten coils of mean diameter 76 mm, pitch 50 mm. If the axial load is 900 N, find the deflection and maximum shear stress. ($E = 207$ GN/m² and $G = 83$ GN/m².)

[13·74 mm. 79·5 MN/m².]

10.19. An open-coiled helical spring has a coil diameter of 70 mm, a wire diameter 8 mm, and a helix angle α of 20°. Find the maximum shearing stress produced in the wire by an axial load of 500 N and the number of coils needed to make the extension under this load at least 30 mm. ($E = 200$ GN/m² and $G = 80$ GN/m².)

[174 MN/m². 7 to the next largest integer.]

10.20. A semi-elliptic leaf spring has five leaves each 7 mm thick and 70 mm wide. What will be the deflection when a load of 7 kN is applied? (Spring span = 700 mm. $E = 207$ GN/m².)

[36·2 mm.]

10.21. A quarter-elliptic spring is to be designed so that it deflects 80 mm under a proof load of 2·2 kN. The material available for the leaves has $E = 200$ GN/m² and will carry a stress that must not exceed 260 MN/m². The thickness must be an integral number of millimetres but the breadth can be any value. The length is 600 mm. Calculate the thickness of leaf and the maximum stress and also suggest suitable values for breadth and number of leaves.

[5 mm. 222 MN/m². $Nb = 1·425$ m, i.e. $N = 13$. $b = 109·6$ mm.]

ANALYSIS OF STRESS

10.22. A semi-elliptic leaf spring 1270 mm long has nine leaves 101·6 mm wide and 9·5 mm thick. What central load would produce a stress of 310 MN/m² and what is the corresponding central deflection? ($E = 207 \times 10^9$ N/m².)
[13·3 kN. 63·6 mm.]

10.23. A thick cylinder with closed ends has an inside diameter of 200 mm and an outside diameter of 400 mm. If the internal pressure is 10^8 N/m² and the external pressure is zero, find the values of radial, hoop, and axial stresses at the inside and at the outside of the walls.

[Inside: $f_r = 10^8$ N/m² compressive,
            $f_\theta = 1\cdot 67 \times 10^8$ N/m² tensile,
            $f_L = 0\cdot 33 \times 10^8$ N/m² tensile.

Outside: $f_r = 0$,
            $f_\theta = 0\cdot 67 \times 10^8$ N/m² tensile,
            $f_L = 0\cdot 33 \times 10^8$ N/m² tensile.]

10.24. A thick cylinder with closed ends has an inside diameter of 100 mm and an outside diameter of 200 mm. What is the maximum pressure it could sustain without yielding at the bore (a) assuming Tresca's yield criterion, and (b) assuming von Mises yield criterion? ($f_{yp} = 400$ MN/m².)
[(a) 150 MN/m². (b) 174 MN/m².]

10.25. Find the ratio of thickness to internal diameter for a tube subjected to internal pressure when the internal pressure is equal to half the greatest circumferential stress. Find the alteration in wall thickness of such a tube 0·2 m internal diameter when the internal pressure is raised to 80 MN/m². (Poisson's ratio = 0·304. $E = 200$ GN/m².)   [London Univ., B.Sc. II]
[0·366. 0·0228.]

10.26. A thick cylinder with open ends has an inner radius r_i and an outer radius r_0. If the internal pressure is p and the external pressure zero, find the maximum pressure that the cylinder can withstand without yielding beginning at the inner wall. Use (a) maximum principal stress, (b) Tresca, and (c) von Mises theories of yielding.

(*Hint.* $f_r = -p$ will be compressive and is the third principal stress f_3; f_θ is tensile and is the maximum principal stress f_1; f_L is the second principal stress f_2 and is zero for an open ended cylinder.)

10.27. Repeat problem 10·25 for a cylinder with closed ends.

10.28. A straight bar of alloy 900 mm long and 13 by 5 mm in section is loaded axially until it buckles. Assuming Euler's formula for a pin-ended strut applies, estimate the maximum central deflection before the material reaches its yield point of 285 MN/m². ($E = 74$ GN/m².)

[London Univ.]

[122·6 mm.]

10.29. Without assuming any formula for critical load, find the maximum axial load that a pin-ended column 1100 mm long, 20 by 5 mm in section and made of steel ($E = 200$ GN/m²), will carry without elastic buckling.

[340 N.]

10.30. A straight bar of steel 21×7 mm in section and 140 mm long has a compressive yield stress of 320 MN/m². Using the Rankine–Gordon formula ($\alpha = 4/30\,000$ for a pin-ended strut), find the critical load (a) when the bar is used as a pin-ended strut, (b) when it is fixed at one end and free at the other, and (c) when it is fixed at both ends.

[(a) 28·75 kN. (b) 13·2 kN. (c) 40·5 kN.]

10.31. A steel cube is made of material whose yield stress in uni-axial tension is known to be 240 MN/m². A tensile stress of 200 MN/m² is applied perpendicular to one pair of faces. What is the maximum compressive stress that can be applied perpendicular to one of the

ANALYSIS OF STRESS

other pairs of faces at the same time without causing failure by yielding? Assume there is no stress in the third direction and that yield points are the same in compression as in tension. Use (a) maximum principal stress theory, (b) Tresca's theory, and (c) von Mises theory.

[(a) 240 MN/m². (b) 40 MN/m². (c) 66·13 MN/m².]

10.32. A shaft is made of steel which has a yield stress, in a tensile test, of 200 MN/m². The shaft is to carry a torque of 2000 N m and is solid and of 50 mm diameter. What is the greatest bending moment that the shaft can carry at the same time as the torque without yielding taking place? Use the two most conservative failure criteria and state which of the two is likely to be the nearer to experimental results.

[Tresca criterion gives $M = 1420$ N m. von Mises criterion gives $M = 1740$ N m.]

10.33. A timber beam 150 mm wide and 300 mm deep is to be reinforced by two steel plates 9·5 mm thick firmly attached one on each side symmetrically about the horizontal centre line of the beam section. If the beam is to carry a concentrated load of 52·5 kN at the centre of a span of 3 m with both ends simply supported and the stress in the timber is not to exceed 8·3 MN/m², find the minimum depth of plate needed. Take $E_s = 20E_T$. [London Univ.]

[227·8 mm.]

10.34. A rectangular wooden beam 100 mm wide by 225 mm deep is reinforced with three plates 75 mm wide by 6·4 mm thick positioned as shown in Fig. 10.10. The beam is simply supported at the ends, has a span of 3·6 m and carries a uniformly distributed load of 3 kN/m. What additional point load can be carried at mid-span if the maximum stress in the steel is not to exceed 62 MN/m²? What will then be the maximum stress in the wood? (Modular ratio = 20 (i.e. E_s/E_t).)
[London Univ.]

[3·9 kN. 2·95 MN/m². ($\bar{y} = 106$ mm. $I_{NA} = 16\cdot91\times10^{-6}$ m⁴.)]

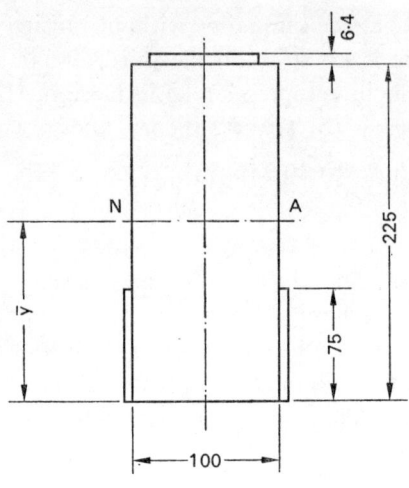

Dimensions in mm
Fig. 10.10.

10.35. A timber beam is made up of two planks each 100 mm wide and 230 mm deep. A recess 150 mm wide and 9·5 mm deep is cut symmetrically about the longitudinal centre line of one face of each plank and a steel plate 150 mm wide and 19 mm thick is placed in the recess and the planks firmly fastened together and to the plate so that the steel is completely enclosed and the resulting compound beam is 200 mm wide and 230 mm deep. Calculate the resisting moment of the section when the maximum bending stress in the timber is 8·3 MN/m². What is then the maximum stress in the steel? ($E_s = 200 \times 10^9$ N/m². $E^T = 12 \cdot 5 \times 10^9$ N/m².) [London Univ.]

[20·8 kN m. 86·8 MN/m².]

10.36. Two bars of different metals each of the same rectangular cross-section 10 mm thick 40 mm wide are firmly fastened together to make a compound bar 20 mm thick. The bar is simply supported over a span of 700 mm with material A uppermost. What is the maximum central point load the bar will support without the stress in material A exceeding 70 MN/m² or the stress in material B exceeding 100 MN/m²? ($E_A/E_B = 3$.)

[1·2 kN (stress in material B is governing: $\bar{y} = 7\cdot5$ mm; $I_{NA} = 4\cdot7\times10^{-8}$ m⁴).]

10.37. A reinforced concrete beam is 305 mm wide and 230 mm deep with tension-reinforcement centred 50 mm from the bottom surface. If the maximum stress in concrete = 4 MN/m² and in steel is 103 MN/m², and both are to be attained simultaneously, calculate the area of reinforcement and the safe uniformly distributed load that can be carried over a simply-supported span of 3 m. ($E_s = 15\ E_c$.)

[392 mm². 4·05 kN/m.]

10.38. A reinforced concrete beam 300 mm wide and 500 mm deep has four reinforcing bars each of 5 cm² area all centred 50 mm above the tension face. If the maximum allowable stress in concrete is 6·55 MN/m² and in the steel is 103 MN/m², find the distance of the neutral axis from the tension face and the resisting moment of the section. ($E_s/E_c = 15$.)

[284 mm. 78 kN m (governed by stress in steel).]

10.39. A reinforced concrete T-beam has the compression flange 140 mm thick and 305 mm wide. The reinforcement is in tension only and consists of 5·9 cm² centred at 305 mm below the top of the compression flange. The web is 152·5 mm wide. Find the moment of resistance if the steel is limited to a stress of 124 MN/m² and the concrete to 5 MN/m². ($E_s/E_c = 16$.)

[14·3 kN m (governed by stress in steel).]

10.40. A concrete beam is to be designed with tension reinforcement to the following specification. Resisting moment = 200 kN m; beam width 500 mm; $E_s/E_c = 15$; $f_s = 110$ MN/m²; $f_c = 4$ MN/m². Find the effective depth and area of reinforcement.

[800 mm. 25·68 cm².]

Appendix I

S.I. Units Used in this Book

Base Units

Quantity	S.I. unit	Multiples and submultiples
Length	m (metre)	mm (1 mm = 10^{-3} m)
Mass	kg (kilogram)	g (1 g = 10^{-3} kg)
Time	s (second)	h, min, ms

Derived Units

Quantity	S.I. unit	Multiples and submultiples
Area	m^2	mm^2 (= 10^{-6} m^2)
		cm^2 (= 10^{-4} m^2)
Volume	m^3	dm^3 (= 10^{-3} m^3)
		mm^3 (= 10^{-9} m^3)
Density	kg/m^3	
Force	N (newton)	MN (= 10^6 N)
		kN (= 10^3 N)
Moment of force	N m	kN m
Pressure, stress	N/m^2	GN/m^2 (= 10^9 N/m^2)
		MN/m^2
		kN/m^2
Energy, work	J (joule)	MJ, kJ
Power	W (watt)	MW, kW
Velocity	m/s	
Acceleration	m/s^2	

Appendix II

Properties of Common Materials

Material	E (GN/m²)	G (GN/m²)	UTS (MN/m²)	Yield Tension (MN/m²)	Yield Shear (MN/m²)	Density (kg/m³)
Steel 0·2%C (hot rolled)	207	83	414	241	145	7820
Steel 0·2%C (cold rolled)	207	83	516	414	248	7820
Steel 0·8%C (hot rolled)	207	83	827	483	290	7820
Steel 0·8%C (oil quenched)	207	83	1241	827	496	7820
Steel (0·4%C, 3·5% Ni)	207	83	1960	1103	662	7820
Copper (cast)	90	42	207	55	—	8900
Copper (hard drawn)	117	42	379	262	159	8900
Brass (60/40)	90	34·5	310	138	—	8295
Aluminium (cast)	69	28	90	62	—	2650

Timber — Density: 35–50 kg/m³
Bending properties: E, 6–12 GN/m²
Working stress, 5–10 MN/m²
Modulus of rupture, (My/I) 40–90 MN/m²

Concrete (1 : 2 : 4 mix) — Cube strength (28 days): 25 MN/m²
Modulus of rupture: 2·4 MN/m
Modular ratio: 15

Index

Arch
  three-pinned   6, 20, 24
  two-pinned   198

Beams   5, 81–111, 146
  bending stresses   83, 94–98, 103–105, 146
  built-in   81, 153, 164, 167, 175–176, 218, 223
  deflection and slope   148, 150–176, 182, 188, 195–196, 198
  of two materials   232, 250, 257–258
  reinforced concrete   233, 251, 259
  shear stresses   83, 94–97, 106–108, 189
Bending moments   82–87, 90, 99–103, 129, 175, 214
Bow's notation   1

Cantilever   17, 20, 81, 84, 100, 102, 161, 169, 171, 174–175, 188, 194, 196, 201
Castigliano
  first theorem   178–180, 182, 187, 196
  second theorem   178, 192
Compound bars   42, 43, 54
Contraflexure, point of   89, 91
Cylinders
  thick   229, 243, 255–256
  thin   37, 47, 56, 74, 143, 229

Eccentric loading   83, 97–99, 108–111
Elastic constants
  definitions   35
  relations between   132–133, 140, 149

Elastic limit   83, 203
Equilibrium equations   4, 6, 7, 67, 84, 89

Flanged couplings   113, 120, 126
Force polygons   1, 12–14
Fracture   71
Frames
  deflection of   185, 199–201
  plane   8–13, 14, 21, 24–26, 28–29, 184–187, 206, 220–221
  portal   187, 198
  redundant   192, 201, 205, 221
  space   14, 31–32
Free-body   1, 2, 16–17, 66–67, 81

Impact loading   38, 45, 57, 58, 191, 198
Inflection, point of   89, 91, 101
Interference fits   57

Lamé equations   229, 243
Limit design   203, 207

Macaulay's method   151, 158, 172, 174
Materials, properties of common   262
Modulus
  bulk   35
  elastic   35, 50, 53
  resilience   58
  rigidity   35, 112
  Young's   35, 50, 53

INDEX

Mohr's circle
  for strain 132–135, 137, 140, 143–145, 147
  for stress 61–63, 68–70, 73, 79, 134–135
Moment-area method 152, 162, 169, 174

Neutral axis 83, 93, 106, 205, 212, 214

Plastic
  bending 204
  collapse 206–208, 218, 223
  deformation 37, 152, 203
  hinge 204–205, 212, 218–219
  stress analysis 202–223
Poisson's ratio 35–36, 39, 132, 136
Principal planes 63, 72, 74–75, 135
Principal stresses 63, 69–77, 79, 122–123, 128, 129, 135, 147

Reaction calculations 4–8, 10, 13, 18–20, 84–86, 100, 187
Resilience 38

Second moment of area 83, 91–92, 106
Section modulus 217, 222
Sections, method of 11, 25
Shafts, composite 117, 125–126
Shear
  deflection 188–190, 201
  force 82, 84, 87, 90, 99–102, 175
  force sign 82
  stress 34, 71, 83, 123, 125–130
    complementary 60, 73, 113
    distribution in beams 83, 94–97, 106–107, 189
    in complex systems 62–64, 72, 74–75

S.I. units 261
Springs
  helical close-coiled 224, 235–238, 252, 253
  helical open-coiled 226, 238–241, 253–254
  leaf 228, 241–242, 254–255
Statics 1–33
Strain 35–36, 39–40, 131–149
  energy 38, 46, 58, 177–201, 225
  gauge 55, 131, 135, 137, 143–147
  gauge rosette 137–138, 144–145
  shear 131
  volumetric 35
Stress
  and strain 34–59, 132, 202
  complex 68, 239
  direct 34, 39, 60, 73, 79, 98
  hoop 37, 48, 56, 143
  shear 24, 60–63, 65, 71
  three-dimensional system 63
  two-dimensional system 60–80
Struts 230–232, 244–248, 256
  Euler's theory 230, 244–246, 256
  Rankine–Gordon formula 231, 256
Superposition 84, 152, 163

Tensile test 36, 53, 54
Theories of failure 232, 248–249, 256–257
Thermal stresses 37, 44, 55–56
Thick cylinders 229, 243, 255–256
Thin cylinders 37, 47, 56, 74, 79, 143
Torsion 72, 112–130, 178–179, 203
  and bending 114, 122, 128–129, 144, 146
  and direct load 79, 122, 128, 144

Yield point 50, 203
Yield stress 50, 202
Young's modulus 35, 50, 53